Analysis and Design
of Sequential Digital Systems

Other titles in Electrical and Electronic Engineering

G. B. Clayton: EXPERIMENTS WITH OPERATIONAL
AMPLIFIERS

G. B. Clayton: LINEAR INTEGRATED CIRCUIT APPLICATIONS

J. C. Cluley: ELECTRONIC EQUIPMENT RELIABILITY

R. F. W. Coates: MODERN COMMUNICATION SYSTEMS

A. R. Daniels: INTRODUCTION TO ELECTRICAL MACHINES

C. W. Davidson: TRANSMISSION LINES FOR COMMUNICATIONS

W. Gosling: A FIRST COURSE IN APPLIED ELECTRONICS

B. A. Gregory: AN INTRODUCTION TO ELECTRICAL
INSTRUMENTATION

Paul A. Lynn: AN INTRODUCTION TO THE ANALYSIS AND
PROCESSING OF SIGNALS

A. G. Martin and F. W. Stephenson: LINEAR MICROELECTRONIC
SYSTEMS

R. G. Meadows: ELECTRICAL COMMUNICATIONS THEORY,
WORKED EXAMPLES AND PROBLEMS

J. E. Parton and S. J. T. Owen: APPLIED ELECTROMAGNETICS

A. Potton: AN INTRODUCTION TO DIGITAL LOGIC

J. T. Wallmark and L. G. Carlstedt: FIELD-EFFECT TRANSISTORS
IN INTEGRATED CIRCUITS

G. Williams: AN INTRODUCTION TO ELECTRICAL CIRCUIT
THEORY

Analysis and Design of Sequential Digital Systems

L. F. Lind

Department of Electrical Engineering Science,
University of Essex

J. C. C. Nelson

Department of Electrical and Electronic Engineering,
University of Leeds

First edition 1977
Reprinted 1979

Published by
THE MACMILLAN PRESS LTD
London and Basingstoke
Associated companies in Delhi Dublin
Hong Kong Johannesburg Lagos Melbourne
New York Singapore and Tokyo

ISBN 0 333 19266 4 (hard cover)
0 333 19267 2 (paper cover)

Set in Monophoto Times by
Doyle Photosetting Ltd, Tullamore, Ireland
Printed by Unwin Brothers Limited,
The Gresham Press, Old Woking, Surrey.

Contents

Preface

This text presents the basic information required for successful design of modern logic systems. The information is presented in a form that should be immediately applicable by practising engineers but that also forms the basis of a comprehensive final-year university or polytechnic course. It will also prove useful to postgraduates in disciplines relying on the application of digital methods, for example, computer science, control engineering and instrumentation.

Certain mathematical theorems relating to Boolean variables are excluded for over-all clarity. The use of these theorems to simplify Boolean expressions has, in large part, been superseded by mapping and tabular techniques. Excessive treatment of the mathematical basis of some of the procedures would tend to obscure the engineering concepts involved.

The theoretical material is illustrated with practical examples wherever possible. The examples can be used for extending the reader's knowledge from the particular to the general, which, in many cases, appears to be a more natural approach than the reverse. Wide use of references is made where this is felt desirable, to free the text from detail not immediately relevant and to provide a basis for further reading.

An important feature of the book is the consideration of logic design from a human point of view. An attempt is made to relate the more formal design techniques to the widely used intuitive or 'cut-and-try' approach. For example, the problem of state assignment is simplified by considering only those possibilities that fit in with a natural (or human) approach to the problem. This consideration has repercussions in all phases of the design procedure. One of the most important results lies in the increased understanding of the operations of the circuit by service personnel. Also, by requiring the solution to look 'natural', the number of possibilities is greatly reduced (sometimes to one), thereby reducing the amount of design time required.

The partitioning of one complex circuit into a number of smaller circuits is also carefully considered. To date there is no automatic procedure for doing this; instead, *ad hoc* solutions are generally used. Guidelines are laid down for accomplishing this partitioning in this book. Again, 'naturalness' is the keynote of the partitioning procedure. A well-partitioned circuit should be relatively simple to test and troubleshoot, which leads to enormous practical advantages in the post-design career of the circuit.

Particular attention is devoted to hazards, races and other phenomena that, although not apparent from a superficial appraisal of a proposed design, can lead to mal-operation. Careful attention to these points, together with the information provided in the final chapter on practical implementation, should lead to the production of reliable designs every time.

The authors would like to thank R. Coleman, Technical Director of Trend Communications Ltd, for his assistance with chapter 6. They would also like to thank Mrs S. Nelson for her accuracy and patience in typing the manuscript.

1

Introduction

Digital systems are not new. The simple 'on–off' signalling techniques used by nineteenth-century telegraphers are one example of an early application of digital techniques. However, for many years development of the various applications of digital principles continued virtually independently; construction of electromechanical telephone exchanges, for example, progressed quite independently of the development of digital computing systems.

Two factors have influenced the present importance of digital-system design as a discipline in its own right. One is the development of design procedures—many of which are discussed in this book—that apply to all digital systems regardless of their particular form of realisation or application. The other is the availability of electronic logic devices in integrated-circuit form, at a price that, a few years ago, would have been regarded as impossibly low. This low price has enabled digital techniques to be used in systems where previously such methods would have been regarded as quite uneconomic.

1.1 What are digital systems?

All methods of specifying quantities are either continuous or discrete. In the former case, for example, the height of a column of liquid in a tube, all heights are possible between zero and some maximum value limited by the length of the tube. A scale might be used to measure the height of the liquid but the liquid itself will move smoothly between the divisions on the scale and, if measurement could be made sufficiently accurate, an infinity of possible heights would be available. Discrete systems, for example, the mileometer (odometer) used on motor cars, have a limited (although perhaps large) number of possible readings or output values. It is normally impossible to specify the distance travelled by the car to better than the nearest whole mile;

if a 'tenths' digit is provided the resolution of the system is improved but the distance can still be specified only to the nearest tenth of a mile.

In general all systems that fall into the discrete category can be regarded as digital, although certain classes of digital system have a more restricted definition. In electrical terminology continuous systems, as opposed to digital ones, are often referred to as *analogue*. This is because most, if not all, physical variables (for example, temperature, pressure, voltage and current) are continuous quantities and it is often convenient to process a voltage or current that is the analogue of some non-electrical quantity. This technique, analogue computing, has been highly developed and is widely used in specialised applications such as control-system simulation. Computers operating in a discrete mode—digital computers—are used in a much wider range of applications.

1.2   How are digital systems realised?

Any system that can be constrained to have a finite number of levels can be regarded as digital. For example, the discs carrying the figures in the mileometer are constrained by mechanical means to have only ten allowed angular positions, as opposed to complete freedom of rotation. The use of ten levels is particularly convenient for systems requiring human involvement. In engineering practice, however, the use of only two levels is particularly attractive. The levels can be widely separated; this allows each level to be severely degraded by deterioration or inadequacy of system components before there is any danger of confusion between the two. For this reason most digital systems operate in a two-level or 'binary' mode, with translation to (or from) ten levels where human intervention is required.

The levels chosen are, very often, merely the presence, or absence, of some physical quantity. In electrical terms this naturally suggests the presence or absence of a voltage (or current); alternatively a circuit, represented by the contacts of a switch or relay, might be 'open' or 'closed'. Hydraulic and pneumatic digital systems are widely used in the control of machinery; here the two levels are represented by the presence or absence of the appropriate pressurised fluid.

Each variable that is permitted to have one of only two possible values is a binary digit (or *bit*). Digital-system design consists essentially of devising an interconnection of processing elements that produces the desired relationship between an input pattern of bits and the required output pattern. In sequential systems, introduced in chapter 3, the output pattern depends not only on the present input pattern but also on previous ones. In view of the use of only two levels, the number of possibilities when acting on a given number of digits can be unambiguously defined. Details are discussed in chapter 2. The basic principles of design discussed in this book apply regardless of the physical form that the binary variables take. However, purely electronic realisation is by far the most common and this book is written primarily from this point of view. Electronic logic circuits are now normally, but not always, realised in integrated-circuit form. Integrated circuits consist of the components required for realisation of a series of digital functions produced on a single 'chip' of silicon that is only a few millimetres square.

1.3 Binary representation of quantities

An appropriate code must be used if quantities are to be satisfactorily represented over a wide range using binary digits. Although by no means the only possibility, pure binary code is very widely used. In this code adjacent digits are given values (or *weights*) that are related by a factor of two, starting with 2^0 ($=1$), just as adjacent digits in decimal numbers are related by a factor of ten. In the decimal system the magnitude of each digit must be specified between zero and nine; in the binary system each digit can be only present (*one*) or absent (*zero*). For example, 1101 represents

$$1 \times 2^3 + 1 \times 2^2 + 0 \times 2^1 + 1 \times 2^0 = 8 + 4 + 0 + 1 = 13$$

By a similar process decimal fractions can be specified as a sum of binary fractions. Note that, because it normally takes a finite time to convert a physical quantity into binary digits, and these digits can change only at a finite rate, digital quantities are discretised not only with respect to magnitude but also with respect to time (that is, they are *sampled*).

Quantities that can be either positive or negative require an additional digit to specify the sign. Although 'sign-and-magnitude' notation is occasionally used, two's complement notation is more common. In this, the magnitude of a negative number is subtracted from 2^n to form the complement, where n is the number of digits in the magnitude part of the number. This approach has the advantage that subtraction may be performed merely by adding the complement of the number to be subtracted.

Although convenient from a processing point of view, pure binary code leads to hardware complexities when a decimal input or display is required. For this reason, systems in which decimal input or output facilities are a significant part of the total system (for example, digital-voltmeters and pocket calculators) often work in a binary-coded decimal (BCD) mode. In this, numbers are represented in decades, but within the decades the ten levels are specified by means of binary code. For example, 0001 0011 1001 represents 139. The price of easy conversion to and from decimal is that more digits are required and some operations, particularly arithmetic, become more complex. Codes having other weightings (for example, 1, 1, 2 and 5) have been used to represent the ten levels within each decade. These are to be avoided since true binary-coded decimal is widely used and any deviation from this can result in incompatible pieces of equipment.

1.4 Serial and parallel data

It is evident from the previous section that transmission of non-trivial numerical information requires several digits. These digits can be transmitted simultaneously (*in parallel*) or sequentially (*serially*). The former has the advantage of speed but requires one transmission path for each digit. Serial transmission requires only one path regardless of the number of digits but restricts the speed of transmission by an amount that, for a given transmission system, is proportional to the number of digits. Parallel transmission would normally be used over short distances (perhaps within a system). For economy over longer distances a serial method would normally be used. Parallel-to-serial and serial-to-parallel conversion is readily achieved (see chapter 3).

1.5 Comparison of analogue and digital systems

It could be inferred easily from section 1.1 that analogue systems are potentially more accurate than digital ones, which have a finite resolution. In practice this is not the case; in all natural processes there is an inevitable randomness, generally described as *noise*, which means that the infinitesimal resolution that is theoretically possible can never be realised as a reliable measurement. Quantities represented in digital form, however, can often be interpreted with no loss of accuracy even in the presence of considerable noise; but when the noise exceeds a certain threshold, errors can be very large.

These concepts can be illustrated by means of the elementary examples of section 1.1. Unfavourable conditions such as poor visibility, vibration and an unstable location for the observer would decrease the accuracy with which the height of liquid in the tube could be measured. On the other hand the figures on a series of dials, such as a mileometer, could be read without loss of accuracy under such conditions. However, if the conditions were to become particularly poor, a 6 (for example) could be mistaken for an 8 or a 9 and relatively large errors would occur.

Digital systems operating properly below the critical level of noise will always provide the same results for a given series of inputs. This can give rise to a false sense of confidence since errors due to sampling and discretisation will always be present on the input data that the system processes.

In general, therefore, analogue systems are less precise than digital ones but are often faster since they can operate at their maximum rate without the need for sampling. Historically they have also been regarded as simpler and cheaper. However, the development of complex digital integrated circuits has swung the balance in favour of digital realisation for many applications.

2

Review of Combinational-logic Techniques

2.1 Logic levels

Any logic device (or indeed, system) can be viewed as a processing unit. Certain inputs are applied, and certain outputs then appear. These inputs and outputs can take many different forms. For example, they can be voltages, currents, pressure, light, etc., or a mixture of these variables. In the majority of logic devices currently available, the inputs and outputs are voltages. For definiteness, all inputs and outputs in the ensuing discussion will be assumed to be voltages unless stated otherwise.

Having agreed to select voltages to represent inputs and outputs, the next step is to specify how voltage waveforms should be used to convey information. The information of concern is assumed to be a binary digit (bit), which can have only one of two possible values. These values can have various meanings. For example, true, false; on, off; *one*, *zero*; 1, 0 are some of the meanings that have been attached to these two values. The voltage waveforms necessary to distinguish between these values could, in principle, be chosen in a great variety of ways. For example, two different sinusoidal frequencies could be used, or two different phases of one frequency (with respect to some reference phase) could be employed. Alternatively, a square wave and a triangular wave could be used, or two rectangular waves with different mark-to-space ratios. The use of two different d.c. levels could also be considered.

The last possibility is very attractive from the design point of view, since it will probably result in a less complicated circuit than the other ideas mentioned. The use of d.c. levels can be related to the well-known switching properties of transistors (which can be manufactured in integrated-circuit form). It is for this reason that the two values of a bit are usually associated with *d.c.* levels, and are referred to as the *logic* levels of the bit.

Strictly speaking, the association of a logic level with a d.c. level is not correct. The logic level in practice must cover a *range* of d.c. levels. It is necessary to have a range in order to cope with the problems of noise pick-up, varying power supply, ageing of components, etc. In transistor–transistor logic, for example, inputs are allowed in the ranges -1.5 to $0.8\,$V (logic level *zero*) and 2.0 to 5.5 V (logic level *one*). The corresponding output ranges are 0 to 0.4 V and 2.4 to 5.0 V. It is seen that both input ranges are somewhat larger than the corresponding output ranges. The minimum difference in ranges is defined to be the *worst-case* noise margin for each logic level. For example, logic level *one* has a worst-case noise margin $2.4 - 2.0\,\text{V} = 400\,\text{mV}$. In this case the same noise margin exists for logic level *zero*.

For both input and output ranges there exists a forbidden band of values. If an input exists in this forbidden band, the output might be anywhere from 0 to 5 V (which includes its forbidden band). Whatever voltage the output assumes, this value could change dramatically if the original device were replaced by another of the same type. For this reason, operation in the forbidden bands is avoided in practice.

There are many textbooks that give detailed design information on logic levels and noise immunity for various logic families. The reader is referred to Motorola (1968, 1973) for more details.

2.2 Gates

The majority of the logic gates that are encountered in practice are of the multiple-input/single-output type. Thus the gate can be considered as an information-*combining* unit. It accepts several input bits of information and then processes these bits to produce only one output bit. In a sense, information is being destroyed (or entropy increased) in the gate. This entropy point of view has been developed by Matheson (1971a, 1971b).

An important feature of electronic gates is that of unidirectional operation. No matter what signals are impressed on a gate output, the input levels normally remain unaffected. This situation is not true, for example, with relay circuits. With these bilateral circuits, backward paths can occur, which might give rise to incorrect operation.

The simplest way to describe the action of a gate is to prepare an input/output table for the gate. Such a table is conventionally referred to as a *truth table*. This table should show the resulting output for all possible input combinations. Inputs will be denoted by $x_1, x_2, \ldots, x_n$, and the output by z. All these variables are binary, and so can assume only the logic levels *zero* or *one* as discussed previously.

x_1	x_2	z
0	0	0
0	1	0
1	0	0
1	1	1

Figure 2.1   *Truth table and symbol for a two-input AND gate.*

As a first example the truth table of a two-input AND gate will be considered. This truth table is shown in figure 2.1, where logic level *one* is indicated by 1, and logic level *zero* is indicated by 0. Note that all possible input combinations are listed in the table. The symbol for this gate is also shown in figure 2.1. The gate symbols used in this book are in accordance with MIL STD 806 B. It is common practice for manufacturers to use this symbol set. In appendix 1 a list of equivalences is given between these symbols and other widely used equivalents.

The action of this gate can also be described algebraically, as follows

$$z = x_1 x_2$$

In this notation 'multiplication' indicates the AND operation. Letting x_1 and x_2 assume definite values and referring to the truth table of figure 2.1, it is seen that

$$0 \cdot 0 = 0$$
$$0 \cdot 1 = 0$$
$$1 \cdot 0 = 0$$
$$1 \cdot 1 = 1$$

which explains why the AND operator is represented by multiplication. It is obvious that $z = x_1 x_2 = x_2 x_1$, which shows that the AND operation is commutative.

The action of the AND gate can be extended to the n inputs $x_1, x_2, \ldots, x_n$. Then $z = x_1 x_2 \ldots x_n$, which indicates that $z = 1$ only when all $x_i = 1$.

x_1	x_2	z
0	0	0
0	1	1
1	0	1
1	1	1

Figure 2.2 *Truth table and symbol for a two-input OR gate.*

The truth table and symbol for the two-input inclusive OR gate are shown in figure 2.2. Algebraically this truth table is represented as

$$z = x_1 + x_2$$

where the plus sign indicates the OR operation. Again, letting x_1 and x_2 assume definite values, the table shows that

$$0 + 0 = 0$$
$$0 + 1 = 1$$
$$1 + 0 = 1$$
$$1 + 1 = 1$$

which agrees with our notion of addition (except for the last line). This gate is called inclusive OR because the last line gives $z = 1$ for both $x_1 = 1$ and $x_2 = 1$. The operation is again commutative, for $z = x_1 + x_2 = x_2 + x_1$. There is another gate, the exclusive OR gate, which has the property $z = 1$ when either but not both inputs are at *one*. This gate excludes or prevents the output becoming *one* when the two inputs are simultaneously at *one*. The exclusive OR operation is also commutative.

In the inclusive OR gate, let n inputs $x_1, x_2, \ldots, x_n$ be applied. Then $z = x_1 + x_2 + \cdots + x_n$, with the result that z is *one* whenever one or more of the inputs are at *one*.

Some simple rules for the 'addition' and 'multiplication' of logic expressions are now given. Recalling that 1 (or *one*) stands for 'on' and that 0 (or *zero*) stands for 'off', it is readily seen that for some logic expression g, $1 \cdot g = g$, $0 \cdot g = 0$, $1 + g = 1$, $0 + g = g$. It also turns out that $(e+f)(g+h) = eg + eh + fg + fh$, as is the case for algebraic multiplication. With a combination of these rules, various results can be obtained. For example, $(x+g)(x+h) = xx + xh + xg + gh = x + x(g+h) + gh = x(1+g+h) + gh = x \cdot 1 + gh = x + gh$. Rather than develop a long list of such specialised formulae, it is quite often easier to work directly with Karnaugh maps, as will be seen subsequently.

Returning to the discussion of gates, the simplest gate of all is the inverter of figure 2.3. This gate can be represented by a circle or a triangle followed by

Figure 2.3   *Truth table and symbol for an inverter gate.*

a circle, and has the property of complementing (or loosely, 'inverting') the input. The output is often indicated by $\bar{x}$, where the bar indicates the complement (or logical inverse) of x. The definition of a logical inverse is that $x + \bar{x} = 1$. From the truth table, it is readily seen that $x\bar{x} = 0$ and $\overline{(\bar{x})} = x$.

The first equation will be central to the Karnaugh map, which is discussed in section 2.3. The last equation will prove to be extremely useful for logic-circuit transformation. In words, this equation states that two (or any even number, for that matter) inverters in series cancel each other out. Conversely, it is possible to 'grow' two inverters on a connection wire without disturbing the operation of a logic circuit.

The inverter can be combined with the AND- and OR-gate outputs to create the NAND and NOR gates, respectively. The N indicates that the output is negated (that is, inverted). The symbols and truth tables for these gates are shown in figure 2.4. In this figure it is seen that if $x_1 = x_2$ both of these gates reduce to inverters.

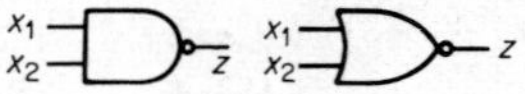

x_1	x_2	z
0	0	1
0	1	1
1	0	1
1	1	0

x_1	x_2	z
0	0	1
0	1	0
1	0	0
1	1	0

Figure 2.4   *Symbols and truth table for two-input NAND and NOR gates.*

Inverters can also be added to the inputs of a gate. Consider what happens when inverters are added to the inputs of the gates of figure 2.4. By referring to the truth tables of this figure, it can be seen that the equivalences of figure 2.5 hold. For example, the inputs $x_1 = 0$, $x_2 = 0$ become inverted ($\bar{x}_1 = 1$, $\bar{x}_2 = 1$), which results in $z = 0$ for both cases in figure 2.4. Other input combinations can be checked in a similar manner. Whenever one of the symbols of figure 2.5 appears in a logic circuit, it can be immediately replaced by its equivalent, if desired.

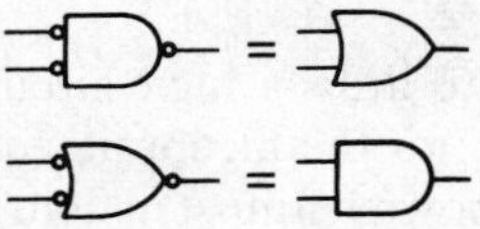

Figure 2.5   *Gate equivalences.*

These results can be used to produce a number of equivalent circuits by using a simple pictorial technique. The method is best illustrated by example.

Example 2.1

Find equivalent circuits for the NAND and NOR gates of figure 2.4.

Solution   Double inverters are grown on each input lead. The equivalences of figure 2.5 are then used. The process is shown pictorially in figure 2.6.

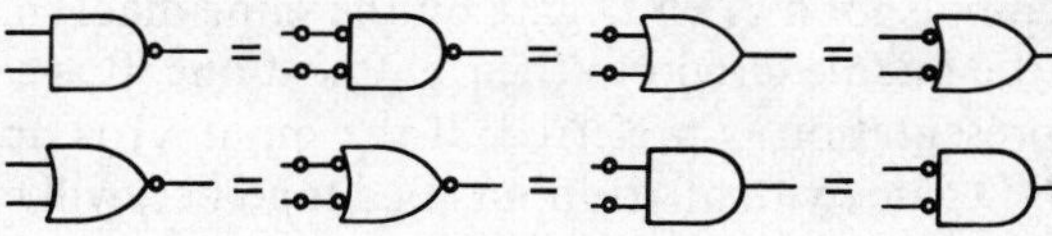

Figure 2.6   *Pictorial solution for example* 2.1.

Example 2.2

Extend the first equivalence of figure 2.5 to a four-input gate.

Solution   The steps of figure 2.7 are used to develop the solution. In this

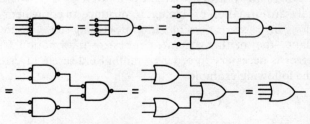

Figure 2.7   *Pictorial solution for example* 2.2.

figure it is understood that, if an inverter is adjacent to a gate, it is formed electronically within the gate itself. If the inverter is elsewhere on a connection wire, it represents a separate electronic entity or, possibly, a hypothetical inverter. This solution can be extended to any number of input leads. A similar result can be obtained for the second equivalence of figure 2.5.

Example 2.3

Realise the function $z = x_1 x_2 \bar{x}_3 + x_4 x_5$ by using NAND gates.

Solution   The first step is to draw a logic circuit (called a *prototype circuit*) which corresponds directly to the algebraic equation for z. The prototype circuit for this example is shown leftmost in figure 2.8. The next step is to grow

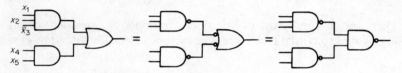

Figure 2.8   *Equivalent-circuit solutions for example* 2.3.

double inverters in such a way that all the gates become NAND gates. From example 2.1 it was found that a NAND gate can be formed either by adding an inverter to the output of an AND gate, or by adding inverters to the inputs of an OR gate. This step is performed in the middle circuit of figure 2.8. In fact, this circuit is the solution to the problem. Many designers prefer not to have two different symbols for a NAND gate on the same diagram, however, and so will convert the middle circuit to the rightmost one. It is a matter of taste as to which representation is preferred. If the input $\bar{x}_3$ is not available, an additional NAND gate (with all its inputs tied together) will be necessary for the inversion of x_3.

The last example illustrated that a prototype circuit consisting of AND and OR gates can be converted to one using only NAND gates. This principle can be extended to *any* prototype circuit consisting of AND, OR, NAND and NOR gates. This prototype circuit can always be transformed into a NAND-gate-only circuit by the inverter-growing process (with discrete inverters provided by degenerate NANDs). This argument can be applied also to the transformation of any prototype circuit to a NOR-gate-only circuit.

The usual transformation problem is to produce an equivalent circuit with as few *discrete* inverters as possible. In a processing sense, discrete inverters are redundant—they neither generate nor destroy information. Usually some trial and error is necessary in order to minimise discrete inverters, as illustrated in the following example.

Example 2.4

Realise the exclusive OR function $z = x_1 \bar{x}_2 + \bar{x}_1 x_2$ with two-input NAND gates using as few gates as possible. The complemented variables $\bar{x}_1$ and $\bar{x}_2$ are not available elsewhere and so must be generated as part of the solution.

Solution   The direct algebraic realisation of z and its subsequent transformation is shown in figure 2.9. Five gates are required, with two of the gates

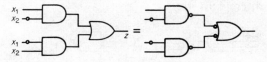

Figure 2.9   *Direct realisation of example* 2.4.

acting as inverters. In searching for a simpler solution, a promising idea would be to use one gate to produce $\bar{x}_1 + \bar{x}_2$ (a NAND gate is equivalent to an OR gate with inverters on the input). The gate is then 'more active' than it would be with both of its inputs connected together. This output, when ANDed with x_1, produces $x_1\bar{x}_2$ (since $x_1\bar{x}_1 = 0$), which is one of the desired output terms. The rest of the design is straightforward, and is shown in figure 2.10. It is seen that this indirect realisation uses only four gates, and so represents an improved solution.

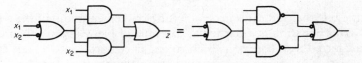

Figure 2.10   *Another solution for example* 2.4.

In attempting to minimise discrete inverters it is generally necessary to use a variety of gate types. For example, sometimes a mixture of NAND and NOR gates will completely eliminate the need for discrete inverters. Also, as a rule it is desirable to use one many-input gate rather than several small ones in minimising the number of gates (and connection wires) needed in the realisation. But in a manufacturing environment it is usually beneficial to reduce the number of gate *types* to a minimum. This procedure eases stocking problems and assembly procedures. There can also be significant price differences between different gates. Therefore, for example, a company might decide to use only two- and four-input NAND gates. Such a policy will usually result in some of the gates being used as simple inverters. This result is not necessarily bad, however. In the implementation of a large system there will usually be some gates left over. These extra gates can then be put to use as inverters. Other criteria can also be used to estimate whether or not a design is optimal. Some of these criteria are discussed in section 5.5.

Occasionally a logic-circuit design requires numerous inverters for the x_i inputs. A general method for dealing with this case will be illustrated with the following example.

Example 2.5

Realise the function $z = \bar{x}_1 \bar{x}_2 \bar{x}_3 + \bar{x}_4 \bar{x}_5$ with AND gates, OR gates and inverters, using as few inverters as possible. The complemented inputs $\bar{x}_1$ are not available and so must be generated within the circuit.

Solution   The transformations of figure 2.11 are used, which result in the saving of four discrete inverters.

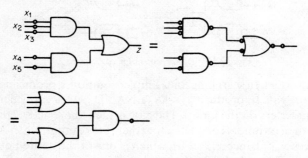

Figure 2.11   *Discrete-inverter reduction for example 2.5.*

The last example led to the equivalence

$$z = \bar{x}_1 \bar{x}_2 \bar{x}_3 + \bar{x}_4 \bar{x}_5 = \overline{(x_1 + x_2 + x_3)(x_4 + x_5)}$$

where the long bar indicates inversion (or complementation) of the ANDed output. This equation is a particular example of De Morgan's theorem. In algebraic form this theorem gives an equivalence between two logic expressions, for which one expression contains complements of all terms and operations of the other expression. It can be written as

$$z = \sum_i P_i$$

$$\bar{z} = \prod_i \bar{P}_i$$

where each P_i is the product (or ANDing together) of input variables (or their complements). For example, let $P_1 = x_1 x_2$ and $P_2 = \bar{x}_1 x_3$. Then $\bar{z} = \overline{(x_1 x_2)(\bar{x}_1 x_3)} = (\bar{x}_1 + \bar{x}_2)(x_1 + \bar{x}_3)$. The natural way to apply De Morgan's theorem for inverter reduction, however, is to use the pictorial procedure of the previous example. This procedure tends to produce fewer errors than pure algebraic manipulation based on De Morgan's theorem.

2.3   The Karnaugh map

The discussion so far has been concerned with gates and the realisation of some output specifications in gate form. Attention is now focused on the minimisation of the specification itself.

2.3.1 Expansion into minterms

The first task is to develop a method that will detect whether two logical expressions are equivalent to each other. For example, the expressions $z_1 = \bar{x}_1 \bar{x}_3 + x_2 x_3 + x_1 x_2 \bar{x}_3$ and $z_2 = x_2 + \bar{x}_1 \bar{x}_2 \bar{x}_3$ are equivalent to each other, but this fact is not obvious from casual inspection of z_1 and z_2. Looking at z_1 and z_2 more closely, it is seen in each case that there are product terms with one or more input variables missing. These missing variables can be recovered by using the fact that $x_i + \bar{x}_i = 1$. Thus, z_1 becomes

$$z_1 = \bar{x}_1 \bar{x}_3 (x_2 + \bar{x}_2) + x_2 x_3 (x_1 + \bar{x}_1) + x_1 x_2 \bar{x}_3$$
$$= \bar{x}_1 x_2 \bar{x}_3 + \bar{x}_1 \bar{x}_2 \bar{x}_3 + x_1 x_2 x_3 + \bar{x}_1 x_2 x_3 + x_1 x_2 \bar{x}_3$$

Applying the method to z_2, it is found that

$$z_2 = x_2 (x_1 + \bar{x}_1)(x_3 + \bar{x}_3) + \bar{x}_1 \bar{x}_2 \bar{x}_3$$
$$= (x_1 x_2 + \bar{x}_1 x_2)(x_3 + \bar{x}_3) + \bar{x}_1 \bar{x}_2 \bar{x}_3$$
$$= x_1 x_2 x_3 + x_1 x_2 \bar{x}_3 + \bar{x}_1 x_2 x_3 + \bar{x}_1 x_2 \bar{x}_3 + \bar{x}_1 \bar{x}_2 \bar{x}_3$$
$$= z_1$$

This expansion appears to be a retrograde step, since simplification of the output expression was the original intention. None the less, the expansion does provide a *standard* form for any output expression (which can be used for equivalence checking). This form contains the sum of a number of product terms, each of which contains *all* the x_i variables. Such a term is referred to as a *minterm*. Two output expressions are said to be the same if each contains the same set of minterms. Once an expression has been put into minterm form, the next task is to find the simplest expression that represents the set of minterms.

2.3.2 Development of the map

Referring to the above example, it is seen that each time a term is multiplied by $(x_i + \bar{x}_i)$, the number of terms increases by one, and each term contains one more variable. Both effects increase the number of gates required in the realisation. Conversely, if two terms can be found that are the same except for a single variable, say $f_1 x_i + f_1 \bar{x}_i$, then these expressions can be combined and the x_i variable eliminated. This procedure is the basis of the Karnaugh-map method.

The Karnaugh map is composed of squares such that each square represents one minterm. An output expression is entered into a Karnaugh map by simply placing 1's in the appropriate minterm squares. A key feature of the map is that minterm squares that differ in only one variable (for example, $x_1 x_2 x_3$ and $x_1 x_2 \bar{x}_3$) are made to lie next to each other. The eye can then spot clusters of 1's which are essential for map simplification. The main problem in the development of a multivariable map is to provide all the adjacencies that are needed while confined to two-dimensional space (for ease of working).

The two-variable map is simple to draw. It is shown in figure 2.12, with three different labelling schemes. The leftmost scheme is the most direct way to indicate minterms. Each minterm is simply the product of a row and a column variable or, in set-theory terms, the intersection of a row strip and column strip. Usually the complemented variables are not labelled, as shown in the centre map. The binary notation scheme is shown at the right. The 0

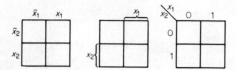

Figure 2.12   *Different forms of a two-variable Karnaugh map.*

indicates $\bar{x}_i$, while the 1 indicates x_i. In figure 2.12, the adjacency between minterms is satisfied. The minterm squares $\bar{x}_1\bar{x}_2$ and x_1x_2 differ by more than one variable change, and so are not considered adjacent (even though their corners touch).

The three-variable map is considered next. In order to meet the adjacency requirements, the natural way to draw this map would be in three-dimensional space. Such a map is drawn at the left in figure 2.13. On this map the x_1 axis is from left to right. Thus the x_1 variable on each block must change from

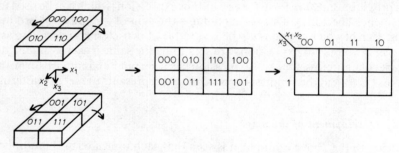

Figure 2.13   *Development of a three-variable Karnaugh map.*

0 to 1 in moving from a left-side block to a right-side block. The same arrangement exists for the other two dimensions. However, it is preferable to represent this map in two-dimensional space. This step can be performed by shifting the blocks 000, 100, 001 and 101 in the figure to the front row. Thus, block 000 is put to the left of 010, and so forth. The shifted block pattern is then viewed from the front, which gives the arrangement in the centre of figure 2.13. This rectangle can be labelled with variables, as shown at the right in the figure. The result is a three-variable Karnaugh map, drawn in two-dimensional space. In order to draw this map, it has been necessary to separate 000 from 100 and 001 from 101. These adjacencies can be retained by imagining the map to be joined together at the left and right edges.

In the construction of a four-variable map the sequence of values for $x_1 x_2$ in figure 2.13 can be repeated along a vertical axis. The result is shown in figure 2.14. In this figure the top and bottom edges are also considered to be adjacent.

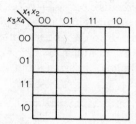

Figure 2.14   *A four-variable Karnaugh map.*

2.3.3   Map entries

When an output expression is given in minterm form the map is readily constructed, by placing 1's in every square that corresponds to a minterm. For example, in the expression $z = \bar{x}_1 x_2 x_3 + x_1 \bar{x}_2 x_3$, the Karnaugh-map entries are shown in figure 2.15. In this map it is apparent that the minterms are not adjacent. Each blank square can be given a 0 entry if desired.

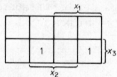

Figure 2.15   *Entries on a three-variable map for $z = \bar{x}_1 x_2 x_3 + x_1 \bar{x}_2 x_3$.*

If the output is not given in minterm form, then the map entries will appear in clusters. For example, let $z = x_2 \bar{x}_3 x_4 + x_1 \bar{x}_2 \bar{x}_4$. Evidently a four-variable map is required. The first term does not contain x_1; therefore, 1's must occur on the Karnaugh map in both the x_1 and $\bar{x}_1$ regions for this term. A similar result holds for the x_3 variable in the second term. The resulting map for this expression is shown in figure 2.16. Loops have been placed around the 1's

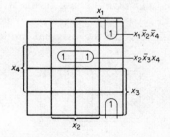

Figure 2.16   *Map entries for $z = x_2 \bar{x}_3 x_4 + x_1 \bar{x}_2 \bar{x}_4$.*

common to each term of z. Note that the 1's in the rightmost column are adjacent since the top and bottom edges of the map are imagined to touch each other. In this example each term with one missing variable gives rise to two 1's on the map.

The next example of interest is $z = \bar{x}_2 \bar{x}_4 + x_1 \bar{x}_3$. The first term gives rise to 1's in all the corners (which are imagined to be adjacent) and the second term gives a square cluster of four 1's in the upper right corner, as shown in figure 2.17. Note in this example that minterm $(x_1 \bar{x}_2 \bar{x}_3 \bar{x}_4)$ is included in more than one term. In general, if a term has k missing input variables, the map will have

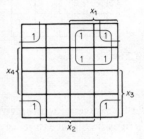

Figure 2.17   *Map entries for* $z = \bar{x}_2 \bar{x}_4 + x_1 \bar{x}_3$.

a cluster of 2^k adjacent 1's. The cluster will be rectangular in form with each side of the rectangle having a number of squares that is a power of two. For example, 1×1, 2×2 and 2×4 are permitted, but 2×3, 1×6 or 2×6 are not.

2.3.4   Map factoring

The process of map factoring involves the grouping of the 1's on a map into rectangles such that the rectangles are as large as possible and such that all 1's are covered at least once. When this has been accomplished the output z can be read from the map in its simplest form. Each rectangle is said to represent a *factor* of the map. This process is the reverse of expanding z into minterm form. In general a rectangle of 1's is called an *implicant* of z. If the rectangle has been made as large as possible (maximum area) it is referred to as a *prime implicant*. If a minterm cell is contained in only one prime implicant, that prime implicant must appear in the output expression. Such an implicant is called an *essential prime implicant*. These terms will be illustrated by the example $z = \bar{x}_1 x_2 \bar{x}_3 + x_1 x_2 \bar{x}_3 + x_1 \bar{x}_2 \bar{x}_3 + x_1 \bar{x}_2 x_3$, which is shown in the map of figure 2.18. The implicants are the original minterms together with the

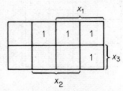

Figure 2.18   *Illustration of implicants, prime implicants, and essential prime implicants.*

terms $x_2\bar{x}_3$, $x_1\bar{x}_3$ and $x_1\bar{x}_2$. The prime implicants are $x_2\bar{x}_3$, $x_1\bar{x}_3$ and $x_1\bar{x}_2$. The essential prime implicants are $x_2\bar{x}_3$ and $x_1\bar{x}_2$. These two implicants include (or cover) all the 1's on the map, with only two factors. The third (redundant) prime implicant $x_1\bar{x}_3$ is not needed in the generation of z.

In general, map factoring consists of finding the prime implicants on a map, and then selecting the minimal (irredundant) set of prime implicants. These two steps are often done simultaneously in map factoring, especially if the number of variables is small. The 'method of attack' is first to factor the 1's that can be combined in only one way. The resulting prime implicants will be essential. The remaining 1's are then factored. In figure 2.18, for example, the 1's representing $\bar{x}_1x_2\bar{x}_3$ and $x_1\bar{x}_2x_3$ can only be combined in one way ($x_2\bar{x}_3$ and $x_1\bar{x}_2$) and so these are essential prime implicants. In this case these essential prime implicants cover all the 1's, and so the map has been completely factored. It would be a mistake to include the implicant $x_1\bar{x}_3$ in z, even though it is the first rectangle spotted on the map.

Referring to figure 2.17 it is seen that 1's can be re-used if necessary in factoring. The goal should always be to make the rectangle as large as possible, regardless of how often 1's are re-used. To illustrate this point, consider the map of figure 2.19. How shall the remaining 1's be factored? If

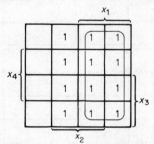

Figure 2.19   *An example of 1's re-use.*

they are lumped together the resulting implicant is $\bar{x}_1x_2$. But this implicant is not prime. The corresponding prime implicant x_2 will include the column x_1x_2, which has already been used in the other prime implicant.

2.3.5   *Many-variable maps*

The Karnaugh map can easily be extended to handle five and six input variables. The five-variable map is shown in figure 2.20. In the middle of the map is a reflection plane, which can be likened to a mirror. In this map minterm A is not only adjacent to B, but is also adjacent to C (by the action of the mirror). Minterms E, F, G and H are all adjacent and thus form a (prime) implicant square.

A reflection plane can be added to the vertical direction, which leads to a six-variable map, as shown in figure 2.21. The figure has sixty-four minterm squares. The number of adjacencies is considerable, and it requires some

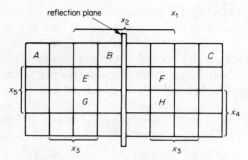

Figure 2.20   *A five-variable Karnaugh map.*

practice to factor the map in an optimal manner. None the less, the afore-mentioned principles are applicable here.

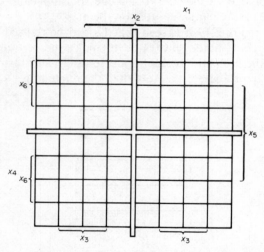

Figure 2.21   *A six-variable Karnaugh map.*

A seven-variable map can be constructed by placing two six-variable maps side by side. An eight-variable map requires four six-variable maps arranged in a square. These higher-order maps are of most use for the case where the number of minterms is reasonably small, and yet large enough for some factorisation to occur. If more than eight input variables are present, the authors have found that it will be necessary either to partition the problem (which is discussed in section 2.4) or to employ computer programs. Examples of such computer programs can be found in Dietmeyer (1971). A comprehensive bibliography of books on logic design is given by Lewin (1968). Finally Potton (1973) gives an expanded treatment of Karnaugh maps and combinational logic.

2.3.6   *Selection of required prime implicants*

The removal of obvious essential prime implicants (in which a 1 can only be combined in one way) has already been discussed. The next problem is to determine which of the remaining prime implicants are to be used.

In general a table can be formed with minterms across the top row and prime implicants down the left-hand side. Such a table is shown in figure 2.22.

| | | required minterms | | | |
		m_1	m_2	m_3	m_4
prime	A	1	1		1
implicants	B		1	1	
	C			1	1

Figure 2.22 *Table used in the selection of essential prime implicants.*

The essential prime implicants can be detected by looking for columns with only a single 1. After these implicants and their associated minterms are removed, it is not immediately obvious which combination of remaining prime implicants will provide cover of the remaining minterms.

An algebraic expression of the complete coverage can be written as follows. Minterm m_1 requires A, minterm m_2 requires $A+B$, and so on. To satisfy *all* minterms, $A(A+B)(B+C)(A+C)$ is required. This expression can be expanded into $(AA+AB)(AB+BC+AC+CC)=(A+AB)\cdot(AB+AC+BC+C)=(A(1+B))(AB+C(A+B+1))=A(AB+C)=AB+AC$, since $(x+1)=1$. The result is that either A and B or A and C can be the required prime implicants. The choice between these two sets is arbitrary. It could well depend on the ease of implementation in relation to other prime implicants. The above procedure was first described by Petrick (1959) and is sometimes referred to as Petrick's method.

The Karnaugh map can be used in this expansion as well. Let $z=A(A+B)\cdot(B+C)(A+C)$. By using De Morgan's theorem (see section 2.2) it is found that $\bar{z}=\bar{A}+\bar{A}\bar{B}+\bar{B}\bar{C}+\bar{A}\bar{C}$. These terms are entered on a three-variable map as shown in figure 2.23. The remaining squares are then given a 0 entry. These

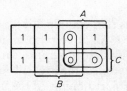

Figure 2.23 *Karnaugh map for the selection of essential prime implicants.*

zeros represent the factors of the complement of $\bar{z}$, that is, the factors of z. Factoring the zeros gives $z=AB+AC$, which is the result of the algebraic expansion above.

Either of these techniques gives *all* combinations of prime implicants that are necessary and sufficient to provide complete coverage.

2.3.7 *Don't cares*

In some designs it is known that one or more input patterns will never occur. If the input is coming from a four-bit decade counter (counting from 0 to 9), for example, it is known that the bit patterns corresponding to the numbers from 10 to 15 should never occur. These unused patterns are commonly referred to as *don't care* conditions. The corresponding minterm squares in the Karnaugh map are marked with ×'s to indicate the don't cares. When factoring the map, these don't cares can be used, *if desired*, for forming larger prime implicants. For example, figure 2.24 shows the map for which $z = 1$ whenever

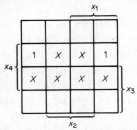

Figure 2.24   *Example of don't cares.*

the count exceeds 7. If the don't cares are not used, then $z = \bar{x}_2\bar{x}_3x_4$. If they are used, however, $z = x_4$. Before minimising a logic expression it is obviously desirable to learn of all the don't cares that will exist in practice. Sometimes it is also desirable to find a minimal expression for all the don't cares. In the unlikely event that one should occur (due to circuit mal-operation), an alarm can be produced. In the above example the don't care alarm would be $z = x_2x_4 + x_3x_4$. Sometimes it is desirable to factor the 0's (or blanks) rather than the 1's. The x's can also be used to simplify this factoring.

2.4 Partitioning

Combinational-logic designs that contain a large number of input variables are occasionally encountered. An attempt to minimise the resulting expressions directly, even with computer programs, can be quite time-consuming and error-prone. An examination of the over-all design requirement will often reveal an inner structure, or partitioning, of the problem. Each block in the partitioning will represent a simplified logic design. This breaking-up of a massive circuit into smaller blocks is useful not only from a design point of view but also for subsequent tasks, such as a description of circuit functions in a manual, and for mal-operation detection. This partitioning is a reflection of man's desire to break up any complex function into block-diagram form, where each block is readily understood. An example at this point will prove instructive.

Example 2.6

Design a two-bit equality detector for the words x_1x_2 and x_3x_4. The output z is *one* only if the corresponding bits in each word are equal. The circuit is to be made with two-input NAND gates.

Solution If a direct solution is attempted, then the first step is to write a complete expression for z, which is $z = \bar{x}_1\bar{x}_2\bar{x}_3\bar{x}_4 + \bar{x}_1x_2\bar{x}_3x_4 + x_1\bar{x}_2x_3\bar{x}_4 + x_1x_2x_3x_4$. If this expression is put on a four-variable Karnaugh map, it will be seen that no factoring is possible. A direct realisation of z requires fifteen two-input NAND gates plus an additional fourteen NAND gates used as inverters.

In seeking a partition for this design, it is worth while to consider the design objective again. On reflection, it is seen that the objective can be met with a single-bit equality block and a block that combines or processes these equalities. The design has therefore been simplified. The design of the single-bit equality block is straightforward. It is specified by the truth table and given by the circuit in figure 2.25. This circuit is the exclusive OR gate of example 2.3,

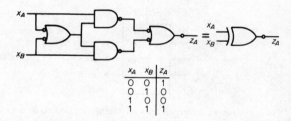

x_A	x_B	z_A
0	0	1
0	1	0
1	0	0
1	1	1

Figure 2.25   *Single-bit equality detector for example 2.6.*

with an inverter on the output. The equality-combining block for this example is simply a two-input AND gate. The complete realisation is given in figure 2.26. It requires nine NAND gates plus three NAND gates used as inverters.

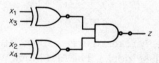

Figure 2.26   *Two-bit equality detector in partitioned form.*

In general it is not known whether a partitioned design will require greater or fewer gates than an unpartitioned one. However, if the partitioned design requires more gates, these extra gates can often be justified by ease of design, ease of understanding and ease of testing. The partitioned design is also readily expandable to handle more digits than the original specification.

2.5   Iterative circuits

The central idea of an iterative design is to perform a partitioning such that all blocks are *the same*, with the possible exception of the first and last blocks. Iterative designs are usually restricted to specifications that contain a great deal of symmetry. An example should clarify these points.

Example 2.7

Design a circuit for the twenty input variables $x_1, x_2, \ldots, x_{20}$ such that z is *one* whenever exactly one of the inputs is at *one*.

Solution   It is clear that a direct approach would result in a massive combinational circuit. In solving this example, it is first imagined that a partitioning block B_i somewhere in the centre of the circuit will have as inputs x_i and the outputs of the previous block B_{i-1}. A list of relevant input and output lines for B_i is formed, such that the output lines are related to the input lines except for the x_i itself. The list for the present example is shown in figure 2.27. It is

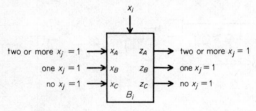

Figure 2.27   *A list of input and output lines for the iterative block B_i in example 2.7.*

seen that a four-input three-output circuit is necessary. In principle a truth table can be produced for each z, and the expression minimised on a Karnaugh map. In this case the necessary expressions can be written down by inspection, as follows

$$z_A = x_A + x_B x_i$$
$$z_B = x_B \bar{x}_i + x_C x_i$$
$$z_C = x_C \bar{x}_i$$

These expressions are easily realised in gate form. Attention must now be paid to the first and possibly last blocks. The first block will reduce to

$$z_A = 0$$
$$z_B = x_1$$
$$z_C = \bar{x}_1$$

(since there are no previous x inputs). The last block can be designed to produce the required output z, which results in the design equation

$$z = x_B \bar{x}_{20} + x_C x_{20}$$

Sometimes it is necessary to treat the first few blocks as special cases. For example, if the requirement for z involves three inputs, blocks B_1 and B_2 will be preliminary stages, having different structures from the iterative block.

The above example illustrates the simplicity of iterative design compared to straightforward design. All of the advantages of partitioning referred to in section 2.3 are enjoyed in an iterative design. A further advantage is that additional blocks can easily be added (or subtracted) if it is decided to change the number of variables at some later stage. One disadvantage is that a change in x_1, for example, will have to 'ripple' through the entire circuit before changing z. If a large number of blocks is being used, the resulting delay could be significant.

2.6 Multiple outputs

In the previous section an iterative design that required three outputs was given. A straightforward realisation of these outputs would involve treating each output variable as independent of the other outputs, the result being a separate map factoring for each output variable. Each z would be only a function of the x_i, and not a function of intermediate outputs of the other z circuits.

The advantages of this procedure are ease of design, ease of circuit understanding and ease of troubleshooting. For example, each z circuit could be put on a separate plug-in card. Repair in this case reduces to card replacement. The disadvantage of the procedure is that more gates are usually required.

The use of common map factors to reduce the number of gates is best illustrated through some examples. In figure 2.28 a three-variable map with

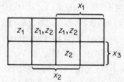

Figure 2.28   *Example of multiple-output simplification.*

the two outputs z_1 and z_2 is shown. It is seen that both z_1 and z_2 require the prime implicant $x_2\bar{x}_3$. This factor can therefore be used in the formation of $z_1 = x_2\bar{x}_3 + \bar{x}_1\bar{x}_3$ and $z_2 = x_2\bar{x}_3 + x_1x_2$, which will reduce the gate count. As a counter example, however, consider the map of figure 2.29. The common area

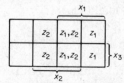

Figure 2.29   *Counter-example of multiple-output simplification.*

$x_1 x_2$ is *not* a prime implicant of either output. In fact the prime implicant of z_1 (that is, x_1) overlaps with the prime implicant x_2 of z_2 to produce the common area. In this case the simplest outputs are $z_1 = x_1$, $z_2 = x_2$, and not $z_1 = x_1 x_2 + x_1$, $z_2 = x_1 x_2 + x_2$.

In general, the use of common factors to simplify multiple-output expressions works best for factors that are common *prime* implicants to two or more outputs. If the common factor is an obvious essential prime implicant to the outputs, so much the better. Usually some trial and error is necessary to see if a real advantage is gained in using common factors for multiple-output circuits. A formal technique for the common-factors approach is given in Hill and Peterson (1968).

2.7 Concluding remarks

In using logic gates a number of practical points should be kept in mind. First of all there is the loading problem. The output of a gate is simply not able to drive an unlimited number of gate inputs. The maximum 'fan-out' (or number of inputs that can be driven) should be established for each family of logic devices (or, in some cases, for each device). The fan-out will represent an upper bound. In some cases it could be necessary to add additional buffer stages to cope with a large fan-out requirement.

The transient problem should also be considered. In the event of sudden changes in all the inputs to a combinational circuit, brief pulses can appear at the output before the output settles down. These pulses are due to the multi-path nature of a combinational circuit, where each path will have a different propagation time. Sometimes these pulses are difficult to see on an oscilloscope, but their presence can be ascertained by their ability to trigger a flip–flop, for example. If these pulses are objectionable, they should be filtered out. The simplest way to accomplish this is by the addition of a small capacitance to the output. More will be said about these transient pulses in chapter 4.

When the output of a logic gate switches from one logic level to the other, there is usually a sudden change in the power-supply current to the gate. This transient, if not controlled, can induce false inputs into other gates for a brief period of time. The popular solution to this problem is to provide many power-supply-decoupling capacitors throughout the circuit. These capacitors may be thought of as small batteries that are able to supply sudden, limited current pulses without much voltage variation. Large electrolytic capacitors should not be used for this purpose, since their inputs can be inductive at high frequencies. These capacitors can be used, however, to reduce substantially any mains interference that might exist.

2.8 Examples

This section contains several examples, both worked and unworked, that illustrate many of the salient points of this chapter. The unworked examples should be capable of solution by readers who have followed the chapter material and the worked examples.

Example 2.8

Realise a twelve-input AND gate by using only two- and four-input NAND gates and two-input NOR gates. The realisation should have as few gates as possible.

Solution The first step is the construction of a prototype circuit. In this circuit only two- and four-input gates should be used. To provide rapid concentration, four-input gates will be used for all inputs, as shown in figure 2.30.

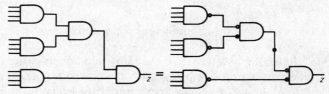

Figure 2.30   *Prototype and final circuit for example* 2.8.

The outputs of these gates can then be combined with two-input gates, or with another four-input gate. (The two-input gates are used in figure 2.30.) The addition of inverters must start at the outputs of the four-input gates, as these gates are to be NANDs. The procedure is then automatic, and yields the circuit on the right in figure 2.30. One discrete inverter is required.

Example 2.9

Find all of the prime implicants in the map in figure 2.31.

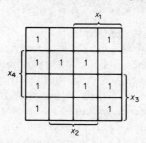

Figure 2.31   *Karnaugh map for example* 2.9.

Solution It is tempting immediately to start drawing rectangles around groupings of 1's. This procedure would undoubtedly locate most of the prime implicants, but the problem statement indicates that all of these implicants are to be found. A more systematic procedure is to go to the first square, identify all prime implicants associated with that square, then go to the second square, and so on. The path through the squares will be arranged to correspond to the way that one reads a page—left to right, top row to bottom row. The result of this procedure is as follows.

minterm square	prime implicants	minterm square	prime implicants
$\bar{x}_1\bar{x}_2\bar{x}_3\bar{x}_4$	$\bar{x}_2\bar{x}_4, \bar{x}_1\bar{x}_2$	$\bar{x}_1\bar{x}_2x_3x_4$	$\bar{x}_1\bar{x}_2, \bar{x}_2x_3x_4$
$x_1\bar{x}_2\bar{x}_3\bar{x}_4$	$\bar{x}_2\bar{x}_4$	$x_1x_2\bar{x}_3x_4$	$x_1x_2x_4, x_1x_3x_4$
$\bar{x}_1\bar{x}_2\bar{x}_3x_4$	$\bar{x}_1\bar{x}_2, \bar{x}_1\bar{x}_3x_4$	$x_1\bar{x}_2x_3x_4$	$x_1x_3x_4, \bar{x}_2x_3x_4, x_1\bar{x}_2x_3$
$\bar{x}_1x_2\bar{x}_3x_4$	$\bar{x}_1\bar{x}_3x_4, x_2\bar{x}_3x_4$	$\bar{x}_1\bar{x}_2x_3\bar{x}_4$	$\bar{x}_1\bar{x}_2, \bar{x}_2\bar{x}_4$
$x_1x_2\bar{x}_3x_4$	$x_2\bar{x}_3x_4, x_1x_2x_4$	$x_1\bar{x}_2x_3\bar{x}_4$	$\bar{x}_2\bar{x}_4, x_1\bar{x}_2x_3$

It is therefore found that eight prime implicants exist.

Example 2.10

Find a minimal set of prime implicants for example 2.9.

Solution   Referring to figure 2.31, the right-side upper corner indicates that $\bar{x}_2\bar{x}_4$ is an essential prime implicant, as only one combination pattern exists for this square. Removal of this prime implicant leads to the map of figure 2.32, in which the removed 1's are represented as don't cares, since these

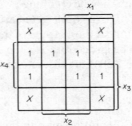

Figure 2.32   *Removal of an essential prime implicant.*

squares can be used again if needed. Forty per cent of the 1's have now been removed! By inspection of the figure it is seen that a minimum of three rectangles (or prime implicants) will be necessary to include all the remaining 1's. One such set is $\bar{x}_1\bar{x}_3x_4, x_1x_2x_4, \bar{x}_2x_3x_4$, and another is $\bar{x}_1\bar{x}_2, x_2\bar{x}_3x_4, x_1x_3x_4$. The second set is simpler and thus represents the 'best' solution. In this example the power of the Karnaugh map as a pictorial display is clearly seen. The map enables man, with his pattern-recognition abilities, to spot the solution (almost) at a glance. The tabular approach—of listing all minterms and prime implicants and then deciding which prime implicants are essential— would be very time-consuming, even for this four-variable problem.

Example 2.11

A long corridor is illuminated by thirty light bulbs having a uniform spacing. By each bulb is a phototransistor that is capable of producing a logic output

x_i. If x_i is *zero*, the ith light bulb is lit. If a bulb failure occurs, x_i goes to *one*. For economy, it has been decided that bulbs will be renewed only when three consecutive bulbs have failed. Design a logic circuit that will detect this condition.

Solution   The nature of the problem suggests an iterative approach. A typical block somewhere in the middle is considered first. This block B_i is shown in figure 2.33. The iterative inputs and outputs that come immediately to mind are 'no failures in a row', 'one failure in a row', 'two failures in a row' and 'three failures in a row' (or 'fault'). Now in passing from one block to the next, there is only one pattern of interest, which progressively goes from one to two to three failures. As far as the iterative outputs are concerned, the output 'no

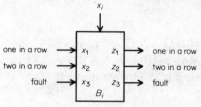

Figure 2.33   *Iterative block for example* 2.11.

failures in a row' is of no interest, and so this variable is dropped from both input and output lines. The remaining iterative inputs and outputs are labelled, and x_i is applied at the top. The logic equations can be written directly from the desired pattern as

$$z_1 = x_i$$
$$z_2 = x_1 x_i$$
$$z_3 = x_3 + x_2 x_i$$

Note that z_2 is *not* $x_1 x_i + x_2$. The problem is not to detect that three bulbs have failed, but that three consecutive bulbs have failed. The above equations are readily realised in gate form.

Example 2.12

Draw a four-variable Karnaugh map for the function $z = x_2 x_4 + x_1 x_2 \bar{x}_3 \bar{x}_4 + \bar{x}_1 \bar{x}_2 \bar{x}_3 x_4 + x_1 \bar{x}_2 x_3 x_4 + \bar{x}_1 x_2 x_3 \bar{x}_4$. Find the essential prime implicants for this function.

Example 2.13

Design a logic circuit that goes to *one* whenever its three-variable input is palindromic ($x_1 = x_3$). Realise this circuit with a minimum number of two-input NOR gates, assuming that the inverted inputs ($\bar{x}_1$, $\bar{x}_2$, and $\bar{x}_3$) are not available externally. (Hint: see example 2.4.)

Example 2.14

Design a combinational circuit that increments a three-bit input number $x_1x_2x_3$ by one. For example, $x_1x_2x_3 = 101$ would result in $z_1z_2z_3 = 110$. The number $x_1x_2x_3 = 111$ should produce $z_1z_2z_3 = 000$. (Hint: make a chart showing all input and output combinations, and then construct a separate map for each z.)

Example 2.15

Tom dates girls who are over 21, have red hair, and are tall, or girls who are tall, not rich, have red hair, and are under 21, or girls who are over 21, rich and tall. Express Tom's preferences in the most economical manner.

Example 2.16

A four-bit binary number $x_1x_2x_3x_4$ is used to represent a decimal digit between 0 and 9. Design a two-input NAND-gate circuit that will produce an output whenever the input is a prime number. Zero is not considered to be prime.

Example 2.17

A logic circuit has an input consisting of a six-bit binary number $x_1x_2 \ldots x_6$. The circuit is to detect when any of the numbers 7, 9, 13, 23, 25, 27, 29, 34, 38, 39, 41, 45, 55, 57, 59 or 61 occur. The numbers 24 and 26 are don't cares. For example, the number 9 is $x_1x_2x_3x_4x_5x_6 = 001001$. Find the essential prime implicants for this circuit.

References

Dietmeyer, D. L., *Logical Design of Digital Systems* (Allyn and Bacon, Boston, 1971) pp. 405–28.

Hill, F. J., and Peterson, G. R., *Introduction to Switching Theory and Logical Design* (Wiley, New York, 1968) pp. 140–8.

Lewin, D., *Logical Design of Switching Systems* (Nelson, London, 1968).

Matheson, W. S., 'PCN Equivalence-class Invariants and Information Quantities', *I.E.E.E. Trans. Comput.*, C-20 (1971a) pp. 691–4.

Matheson, W. S., 'Recognition of Monotonic and Unate Cascade Realizable Functions Using an Informational Model of Switching Circuits', *I.E.E.E. Trans. Comput.*, C-20 (1971b) p. 1214.

Motorola, *Analysis and Design of Integrated Circuits* (McGraw-Hill, New York, 1968).

Motorola, *McMOS Handbook* (Motorola Inc., Phoenix, Ariz., 1973).

Petrick, S. R., *On the Minimisation of Boolean Functions*, Proc. Symp. Switching Theory, ICIP, Paris (1959).

Potton, A., *An Introduction to Digital Logic* (Macmillan, London and Basingstoke, 1973).

3

Introduction to Sequential Systems

3.1 Fundamental concepts

3.1.1 *Defining features of sequential systems*

The systems that have been discussed so far have been purely combinational. The fundamental property of such systems is that, at all times, all of the outputs are determined solely by the present instantaneous values of all the inputs. Of course, in a practical system, there will be some delay between an input change and the resulting changes in one or more of the outputs; but, when equilibrium is reached, a knowledge of all the inputs, together with the logical structure of the system itself, enables all of the outputs to be determined uniquely. In contrast, the outputs of sequential systems are determined not only by the present values of the inputs but also by their previous values.

Clearly, if the outputs are to depend on previous values of the inputs, the system must contain some elements that are capable of *storing* digital information. It is the presence of storage devices, of whatever kind, that distinguishes sequential systems from purely combinational ones.

3.1.2 *Combinational and sequential interpretation of the problem specification*

In some cases a system specification is inherently either combinational or sequential and can be realised only with a system of the appropriate type. Frequently, however, the specification can be interpreted by the designer either in terms of a combinational system or a sequential one. The former approach often has the advantage of speed whereas the latter can be more economical to realise.

Consider, for example, the problem of comparing two binary numbers (N

and M). Three outputs are required; one (z_1) is to be logical *one* when N exceeds M, the second (z_2) is to be logical *one* when M exceeds N and the third (z_3) is to be logical *one* when N and M are equal. This problem is clearly amenable to a purely combinational solution. However, even if N and M are of only four bits each, listing all possible input combinations produces a truth table with $2^4 \times 2^4 = 256$ rows and the minimisation problem is an eight-variable one.

Problems of this kind are often more amenable to an iterative approach, as discussed in chapter 2. It must be emphasised that this approach does not involve storage and, therefore, strictly falls into the category of combinational systems. But, since many of the concepts of iterative systems have their counterparts in truly sequential ones, a brief review of the iterative approach of section 2.5 would not be inappropriate here.

The direct combinational approach makes a simultaneous comparison of all the digits of both numbers. Using an iterative approach, the most significant digits of each number are compared first in a relatively simple circuit. There are four possible combinations of these two binary digits, of which three groups are relevant to the present problem

(a) the most significant digit of N exceeds that of M;
(b) the most significant digit of M exceeds that of N; and
(c) the most significant digits are equal.

If either case (a) or case (b) applies, the relative magnitude of N and M is immediately determined and the remaining digits of the two numbers cannot influence this. If case (c) applies, the relative magnitude of N and M will be determined by the next (less significant) pair of digits unless they are also equal, in which case the third pair must be examined, and so on.

The advantage of this approach is that the relatively simple circuit that compares one pair of digits can be repeated for each subsequent pair and, if a design is to be modified to accept numbers containing more digits, it is merely necessary to add more 'cells' consisting of this basic circuit. As there are three distinct pieces of information to be handed on from one cell to the next, two connections will be required giving one redundant combination. One possible arrangement is given in figure 3.1a. The variables z_1 and z_2 that 'hand on' information concerning the relative magnitudes of previous digits are sometimes called the *state variables*. In this particular example, $z_1z_2 = 00$ is chosen to indicate that all previous pairs of digits examined have been equal to each other, $z_1z_2 = 01$ indicates that M has already exceeded N and $z_1z_2 = 10$ indicates that N has exceeded M. Each cell produces a pair of updated state variables, in accordance with the truth table of figure 3.1b, which are handed on to the next cell. The required outputs, as defined previously, are obtained from the last cell and $z_3 = \bar{z}_1\bar{z}_2$. The detailed design of one cell will be left as an exercise for the reader.

The iterative approach produces a system that is purely combinational since no storage elements have been introduced. In this sense, the use of the

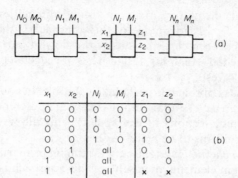

x_1	x_2	N_i	M_i	z_1	z_2
0	0	0	0	0	0
0	0	1	1	0	0
0	0	0	1	0	1
0	0	1	0	1	0
0	1	all		0	1
1	0	all		1	0
1	1	all		x	x

(b)

Figure 3.1 *Iterative approach to the comparison of two binary numbers. (a) Interconnection of cells. (b) Truth table for one cell.*

term 'state variable' to describe the intermediate signals is misleading since this term is widely used in engineering and physics in relation to stored quantities, particularly energies (see, for example, Derusso *et al.*, 1967).

The final approach to the comparator problem is a truly sequential one. A single cell, having similar properties to the one discussed previously, is used to compare each pair of digits in turn, starting with the most significant. The results of each comparison must be stored in suitable devices and these stored values will in turn become inputs for subsequent comparisons. The process continues until all pairs of digits have been compared; details will be discussed later (example 3.5) when the required circuit elements have been considered.

The binary comparator can, therefore, be realised in any of the three forms discussed above. The direct combinational approach would be the fastest but would require the most hardware. The iterative system may or may not require less gates but the response time could be longer in view of the increased number of gate delays involved. The sequential system would be the slowest of all because each pair of bits is compared in turn but could well be the cheapest to implement—a good example of engineering compromise!

3.1.3 *Synchronous and asychronous approaches to sequential-system design*

Sequential systems may broadly be classified as either *synchronous* or *asynchronous*. *Synchronous* systems (discussed in detail in chapter 5) are provided with circuits that are responsible for timing the operation of the remaining sections of the system.

In a simple realisation, the timing circuit consists merely of a square-wave oscillator, or 'clock-pulse generator', and all the storage devices are constrained to change state only when the clock pulse changes from *zero* to *one* or vice versa. The resulting simultaneous, or synchronous, change in the stored values is extremely helpful in avoiding timing difficulties, particularly in complex systems. The disadvantage of this approach is that sufficient time must be allowed between clock pulses for the slowest part of the system to respond and hence the remaining, faster, parts are effectively slowed down to the speed of the slowest.

The synchronous approach is particularly flexible. In complex systems the timing oscillator can have several outputs—each at a different relative phase—that operate distinct sections of the system. This allows many otherwise difficult timing problems to be readily solved.

In contrast, *asynchronous* systems rely for timing primarily on the inherent delays of the system elements themselves. This approach is clearly very fast since a given element in the system begins to respond immediately on receipt of an input change. The associated disadvantage is that, since many elements in the system can be changing state at different times, spurious intermediate values may arise, which cause mal-operation of other parts of the system. For example, consider a stored binary value seven (0111), which is to be changed to eight (1000). In a synchronous system this change would occur virtually simultaneously whereas in an asynchronous system intermediate values of 0110, 0100 and 0000 would be passed through as the storage devices change in turn (from right to left in this case). Consequently any subsequent parts of the system that respond to binary six, four or zero could be incorrectly excited. This problem and related effects, usually referred to as 'races' and 'hazards', are further discussed in chapter 4 and their various manifestations are shown to impose severe restrictions on asynchronous-system design.

3.2   Storage devices

3.2.1   Feedback and storage

From section 3.1.1 it follows that sequential systems consist essentially of combinational logic (the design of which is discussed in chapter 2) and storage devices.

Digital storage devices can be broadly classified as volatile and non-volatile. In the former class, the stored information is lost when the power supply is switched off and must be rewritten on restarting. Non-volatile stores, in contrast, can hold information indefinitely in the absence of power and are of considerable importance in computing systems. The many varieties of magnetic storage systems have this property. In this book, attention will be restricted to storage devices that can be realised by interconnecting logic gates. Such circuits have become a fundamental part of the classical theory of sequential-circuit design and are of increasing practical importance since memories of this kind are now readily realisable in large-scale integrated-circuit form. Memories of this kind are normally volatile.

Purely combinational systems do not make use of feedback. By applying feedback, an array of gates can be given the property of storage; one bit can be stored for each feedback path. Consider the simple feedback circuit shown in figure 3.2a. Assume that inputs S and R are both *zero* and, for the sake of

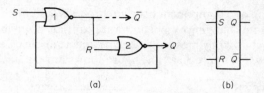

(a)   (b)

Figure 3.2   (a) *Application of feedback to two NOR gates.* (b) *Symbol for the RS flip–flop.*

argument, that the output of NOR gate 1 is also *zero*. Consequently, since both its inputs are *zero*, the output of NOR gate 2 (*Q*) will be *one*. Inputs of *zero* and *one* to gate 1 are entirely consistent with the initial assumed value of *zero* for the output of this gate. The two gates can apparently remain indefinitely in this state, with $Q = 1$ (and $\overline{Q} = 0$), unless some change is forced on the circuit. If an initial value of *one* had been postulated for the output of gate 1, an equally consistent state of affairs would exist with $Q = 0$ (and $\overline{Q} = 1$). Therefore the two cross-coupled NOR gates can store either a logic *zero* or a logic *one*.

The inputs R and S (previously held at *zero*) can be used to change the value of the stored digit. If S is made equal to *one*, $\overline{Q}$ will be forced to zero and, provided R remains at zero, Q will become *one*. Provided S remains at *one* long enough for Q to assume the value *one* (that is, for at least the delay time of the two gates), the second input to gate 1 will become *one* and the stored value Q will remain at *one* even when S returns to *zero*. Further pulsing of S to *one* will have no further effect on the state of the circuit. Conversely, setting the value of R to *one* changes the stored value Q to *zero*. Notice that R and S must not be allowed to become *one* simultaneously. If this were to happen, Q and $\overline{Q}$—which are normally logical inverses of each other—would both be forced to zero and the final state of the circuit would depend on which of R and S returns to zero first.

Although one of the simplest possible examples of feedback, the circuit of figure 3.2a is very widely used. Under the general heading of *flip–flops*, individual circuits of this type are normally identified by the available inputs. The output is the stored quantity Q (and its inverse $\overline{Q}$) and Q may be SET to *one* (hence S) or RESET to *zero* (hence R) and the circuit—which may be symbolised as shown in figure 3.2b—is an *RS flip–flop*. (The 'set' input is sometimes referred to as a 'preset' input and the 'reset' function is also known as the 'clear' or 'preclear' function.)

3.2.2   Clocked and unclocked flip–flops

The basic *RS* flip–flop described in the previous section is *unclocked*. By this it is meant that a change in the state of the flip–flop starts as soon as the appropriate input (R or S) changes and is completed in a time related to the inherent delay of the flip–flop. This is precisely what is required for asynchronous operation.

For operation as a component of a synchronous system, the flip–flop must be prevented from changing except at the appropriate point in the clock-pulse cycle. This is achieved by means of *clocking* and a simple clocked form of RS flip–flop is shown in figure 3.3a. The figure shows how an unclocked flip–flop

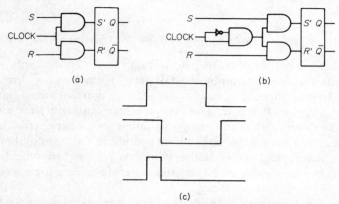

(a)   (b)

(c)

Figure 3.3   (a) *Derivation of the clocked RS flip–flop.* (b) *Simple edge-triggered flip–flop.* (c) *Short trigger pulse generation due to inverter delay.*

(with its inputs renamed R' and S' to avoid ambiguity) may be converted to a clocked one by means of two AND gates. The new S and R inputs can influence the flip–flop only when CLOCK is logic *one*. The appropriate values of S and R must be established in advance and the flip–flop will then change state, in this case, on a *zero*-to-*one* transition of the clock pulse. The clock should then return to *zero* before S or R are allowed to change. More details concerning timing requirements are given in chapter 5.

Although clocked flip–flops of this form (sometimes also referred to as 'bistable latches') are widely used, they have one serious limitation. As long as the clock signal remains at *one*, the S and R inputs of the basic flip–flop remain connected to S and R. In a general system (see chapter 5), the S and R input signals can, in turn, depend on the flip–flop outputs. Consequently, in order to prevent 'new' values of the flip–flop outputs from influencing the input signals and hence the state of the flip–flops, some form of isolation must be introduced. One simple way of introducing this isolation is to make the clock pulse of extremely short duration. It will then have returned to *zero* before the 'new' flip–flop output signals have penetrated the remainder of the system and returned to the flip–flop inputs. This approach is clearly not entirely satisfactory since some 'tuning' is required; the pulse must be long enough to trigger the slowest flip–flop in the system but shorter than the shortest delay in any feedback paths that may be present.

Two distinct approaches have been used successfully to combat this problem.

(1) Edge-triggering: an internally generated short trigger pulse is derived from one edge of the clock pulse.
(2) Master–slave operation: the flip–flop responds to the S or R input on the rising edge of the clock pulse but does not pass on the resulting change of state (if any) to its output until the falling edge of the pulse. The required isolation is, therefore, automatically introduced regardless of the duration of the clock pulse.

The operating principle of an edge-triggered flip–flop is typified by figure 3.3b. There are many detail variations in the circuits actually used to realise such flip–flops in integrated form. The inherent delay—in this case, that of the single inverter—is used to generate an extremely short trigger pulse. As the pulse width will be of the same order as the propagation delay of the single inverter, it is reasonable to assume that this will be less than the delay in any part of the remaining sub-system. Figure 3.3c shows how the short pulse is generated; the upper waveform is a single typical clock pulse. The central waveform is the output of the inverter, where a short propagation delay has been included, and the lower waveform shows the output of the AND gate, which is a short pulse as required (delay in the AND gate has been ignored since it affects equally the clock pulse and its inverse). The pulse width may be increased by including more inverters, provided the total number is odd. This explanation has assumed that the rise time of the clock pulse is very much less than the delay of the inverter; if this were not the case, the short pulse would not be generated properly. For this reason, manufacturers' data sheets normally specify a maximun rise time for the clock input to edge-triggered devices.

The master–slave flip–flop (figure 3.4) does not share this restriction and

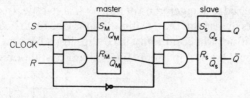

Figure 3.4   *Master–slave RS flip–flop.*

will trigger on a slowly varying waveform—for example, a sine wave. As the clock input changes from *zero* to *one*, the master flip–flop will be set if $S = 1$, reset if $R = 1$ and unchanged if R and S are both zero, exactly as for the clocked flip–flop previously considered. However, the only available outputs from the device (Q and $\bar{Q}$) are derived from the slave flip–flop, whose outputs cannot have changed because its inputs (S_s and R_s) are blocked by the inverted clock signal and the AND gates (a race condition exists here; see the discussion in section 4.3). When the clock signal eventually returns to zero, the master inputs (S_M and R_M) will be blocked from any changes that may occur in S and R, while Q will then, and only then, take on the value of Q_M since the AND gates feeding S_s and R_s are now opened by the inverted clock pulse. To summarise, the master flip–flop takes on the value appropriate to S and R when the clock pulse is at *one* but any resulting change is transmitted to the output of the device (Q) only when the clock pulse returns to *zero*, at which time the inputs S and R become incapable of affecting the master flip–flop. (Some devices operate with a clock signal that is the inverse of that assumed here; in this case the words *one* and *zero* should be interchanged.)

3.2.3   Other types of flip–flop

The RS flip–flop may be regarded as the basic type from which many variations have been developed. The behaviour of the RS flip–flop may be summarised by the table of figure 3.5a, where Q^- is the previous value of the flip–

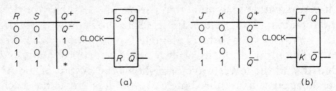

Figure 3.5   *Truth tables and symbols. (a) Clocked RS flip–flop. (b) JK flip–flop.*

flop output and Q^+ is its new value. The prohibited case $R = S = 1$ is indicated by a star and avoidance of this condition imposes a restriction on the system designer. In order to avoid this restriction, the JK flip–flop has been developed and is very widely used. Figure 3.5b shows that operation is exactly the same as for the RS device except that the case $J = K = 1$ is now permitted and always causes a change of state of the flip–flop. Although the RS type may be either clocked or unclocked, the JK type normally exists only in clocked form and can be either edge-triggered or of master–slave type. In addition to being widely used in its own right, the JK flip–flop may be regarded as a general type of which most, if not all, commonly used flip–flops become special cases.

The trigger (or T) flip–flop (figure 3.6a) changes state on every clock pulse

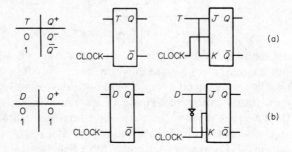

Figure 3.6   *Truth table, symbol and derivation from JK flip–flop. (a) T flip–flop. (b) D flip–flop.*

when T is *one* and remains unchanged when T is *zero*. It can be regarded as, although not necessarily constructed as, a JK flip–flop in which J and K are connected to provide a single input and therefore must always have the same value (that is, only the top and bottom rows of figure 3.5b are accessible). The T flip–flop has one serious shortcoming; in operation, T is made equal to *one* whenever a change in the stored digit is required, whether this be a change from *zero* to *one* or vice versa. This means that if the flip–flop should acquire the wrong state, due perhaps to an interference pulse, changes will continue to occur at the appropriate time but the flip–flop will always be in the wrong state—until a further interference pulse puts it right again!

A frequently used type of flip–flop is the D type where D stands for either 'data', as one bit of data can be stored, or 'delay', since a digit can be delayed by one clock interval. Details are given in figure 3.6b, where it can be seen that the state of the flip–flop after the clock pulse (Q^+) is simply the value of D before it. This flip–flop can be regarded as, but again not necessarily constructed as, a JK flip–flop in which J and K are always constrained to be the logical inverse of each other. Hence $J = D$ but $K = \bar{D}$, and only the second and third rows of figure 3.5b are accessible. A related, but functionally different, device is the *latch*. The output of this follows the input continuously when the clock has one value and stores the input value that exists at the instant when the clock changes to its other value.

The remaining types of flip–flop that are normally encountered are either minor variations of those already described or a combination of the properties of more than one flip–flop in a single device. For example, clocked JK flip–flops often have asynchronous SET and RESET inputs that over-ride the synchronously controlled J and K inputs and are used, primarily, for setting (often asynchronously) the initial state of a synchronous system.

Internal AND gates are often used in JK flip–flops to provide multiple J and K inputs; in such a case the effective value of J might, for example, be the logical AND of inputs J_1, J_2 and J_3. Finally, some or all of the available inputs to the device may be the logical inverse of those discussed previously. For example, if a flip–flop is constructed by means of two cross-coupled two-input NAND gates (as opposed to NOR gates considered previously), the SET and RESET inputs must normally be held at *one* for correct operation and changed to *zero* only to perform the set or reset operation. Flip–flops of this type are often loosely described as RS but should strictly be referred to as $\bar{R}\bar{S}$; normal RS operations could be obtained, of course, by logically inverting both input signals.

Sometimes the K (but not the J) input is inverted. This means that an inverter is required to obtain true JK operation but D-type operation could be obtained by connecting J and $\bar{K}$ directly. No widely accepted convention concerning the significance of CLOCK and $\overline{\text{CLOCK}}$ seems to have emerged. This means that manufacturers' data sheets should always be consulted in order to establish whether the device under consideration changes state on the positive- or negative-going edge of the clock pulse.

3.3   Sequential sub-systems

3.3.1   Basic sequential circuits

In the design of sequential systems several basic configurations recur frequently, occasionally as simple systems in their own right, but more usually as sub-systems within a more complex configuration. For this reason various classes of sub-system have been identified, some of which will be discussed in later sections. In some cases designs can be accomplished merely by interconnecting appropriate sub-systems; examples will be discussed later.

3.3.2   Counters

Possibly one of the first forms of sequential digital system, counters are of wide application. The requirement is to produce a digital output that, in any required code, is a measure of the total number of input (clock) pulses that have occurred since some selected starting point. The specification of a given counter is defined by the code in which counting is performed and the mode of operation (synchronous or asynchronous). Of course, there will be an upper limit to the count, at which point the counter returns to its initial state and starts counting again.

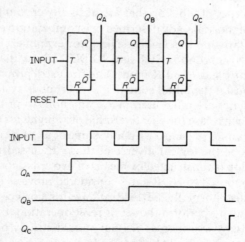

Figure 3.7   *Asynchronous three-bit binary counter and its waveforms.*

Figure 3.7 shows a three-bit asynchronous binary counter that counts from 0 to 7 and uses (unclocked) T flip–flops, which, in this case, trigger on the negative-going edge of the input pulse. The first flip–flop will trigger on every input pulse, whereas the second will trigger only on negative-going edges of Q_A and so on. It can be seen that, at any instant, $Q_C Q_B Q_A$ represents a binary number indicating the total number of input pulses that have occurred since the counter was set to zero by means of the reset inputs. (If the total number of pulses should exceed seven, the counter will recycle and indicate the total number minus an appropriate multiple of eight.) The waveform diagrams also show the cumulative delays that arise from the asynchronous nature of the interconnections and result in the outputs giving a spurious count during transitions. These waveforms are idealised. An actual waveform could have in addition ringing, spikes, d.c. pedestals, and mains hum!

Counters that operate downwards are sometimes required. It can easily be verified that the counter under discussion can be made to operate in the reverse sequence by connecting the T inputs of the second and third flip–flops to the $\overline{Q}$ (instead of Q) outputs of their predecessors. Alternatively, this counter can be formed by using the outputs $\overline{Q}_C$, $\overline{Q}_B$ and $\overline{Q}_A$ of figure 3.7.

To obtain synchronous operation, all flip–flops must be triggered directly from the input as shown in figure 3.8 (where the asynchronous reset connections have been omitted for clarity). Clearly, additional inputs are required to

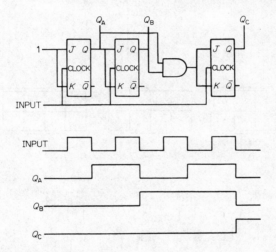

Figure 3.8 *Synchronous three-bit binary counter and its waveforms.*

prevent all the flip–flops changing state on every clock pulse. In the arrangement shown in the figure, Q_A changes on the negative edge of every clock pulse (since this flip–flop has $JK = 11$) but Q_B changes only when $Q_A = 1$ and similarly a change in Q_C requires Q_A and $Q_B = 1$. The waveforms show that this mode of connection produces the required binary sequence; a more formal derivation may be made using the techniques employed in example 3.1. The waveforms shown in figure 3.8 make no allowance for flip–flop operating delays. In reality the flip–flop output changes will be delayed with respect to the falling edge of the clock pulse by one, and only one, flip–flop delay (unlike the case of figure 3.7), but this delay will, nevertheless, be susceptible to normal device-production tolerances.

A wide variety of counters, both synchronous and asynchronous, are available as medium-scale integrated circuits. Any required sequence may be readily realised; the techniques developed in chapter 4 may be used for asynchronous circuits (a simple example is considered in example 3.9) and the following example illustrates the approach required for synchronous operation.

Example 3.1

Design a three-bit synchronous counter that operates in the Gray code of the flow table in figure 3.9a and uses D-type flip–flops.

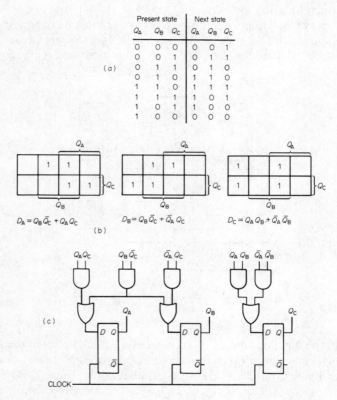

Present state			Next state		
Q_A	Q_B	Q_C	Q_A	Q_B	Q_C
0	0	0	0	0	1
0	0	1	0	1	1
0	1	1	0	1	0
0	1	0	1	1	0
1	1	0	1	1	1
1	1	1	1	0	1
1	0	1	1	0	0
1	0	0	0	0	0

(a)

$$D_A = Q_B \bar{Q}_C + Q_A Q_C$$

$$D_B = Q_B \bar{Q}_C + \bar{Q}_A Q_C$$

$$D_C = Q_A Q_B + \bar{Q}_A \bar{Q}_B$$

(b)

(c)

Figure 3.9   *(a) State table, (b) flip–flop input maps and (c) circuit realisation for example* 3.1.

Solution   A Gray-code sequence (in which each state change involves the change of only one bit) is shown in the 'present state' column of figure 3.9a, where Q_A, Q_B and Q_C represent the outputs of the three flip–flops to be used in the counter. For convenience the required 'next states' are recorded in an adjacent column. The D flip–flop (section 3.2.3) requires an input (D) that is *one* when, and only when, the required next output value is *one*. Hence, it can be seen that flip–flop A requires a D input (D_A) that is *one* when $Q_A Q_B Q_C = 010$, 110, 111 and 101. This information is most conveniently recorded on a Karnaugh map as shown in figure 3.9b, which also includes maps for D_B and D_C. Minimal input functions may now be read off and implemented as shown in figure 3.9c.

3.3.3 Counters having more than one operating sequence

The counters discussed so far have only one specified operating sequence. In general, a counting system may have many possible sequences; the one required for a specific purpose can be selected by means of control inputs. A frequently encountered example is the reversible counter, which operates, in a given code, either upwards or downwards depending on the value of a 'count direction' input signal (alternatively, two clock inputs may be provided, one for 'up' and one for 'down'; no control input is then required).

For all types of multiple-sequence counter, it is essential to ensure that the sequence control signals do not change at the instant when an input pulse occurs (or that both clock inputs are not pulsed simultaneously for those types that have two pulse inputs). Otherwise the required operating sequence cannot be specified unambiguously. The design of reversible counters is illustrated by means of the following example.

Example 3.2

Develop the Gray-code counter of example 3.1 to provide an up/down counting capability. A control signal UP is *one* when upward counting is required.

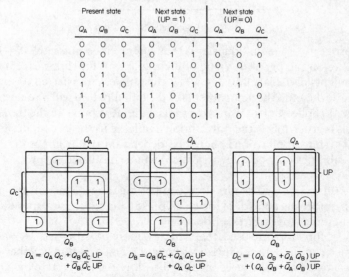

Present state			Next state (UP = 1)			Next state (UP = 0)		
Q_A	Q_B	Q_C	Q_A	Q_B	Q_C	Q_A	Q_B	Q_C
0	0	0	0	0	1	1	0	0
0	0	1	0	1	1	0	0	0
0	1	1	0	1	0	0	0	1
0	1	0	1	1	0	0	1	1
1	1	0	1	1	1	0	1	0
1	1	1	1	0	1	1	1	0
1	0	1	1	0	0	1	1	1
1	0	0	0	0	0·	1	0	1

$$D_A = Q_A Q_C + Q_B \bar{Q}_C \, UP + \bar{Q}_B \bar{Q}_C \, \overline{UP}$$

$$D_B = Q_B \bar{Q}_C + \bar{Q}_A Q_C \, UP + Q_A Q_C \, \overline{UP}$$

$$D_C = (Q_A Q_B + \bar{Q}_A \bar{Q}_B) \, UP + (Q_A \bar{Q}_B + \bar{Q}_A Q_B) \, \overline{UP}$$

Figure 3.10   *(a) State-transition table and (b) flip–flop input maps for example 3.2.*

Solution   The modified state-transition table is shown in figure 3.10a, where $UP=0$ implies downward counting, of course. Three flip–flops are still required but the input signals (D_A, D_B and D_C) become functions of four variables since the required next states are now defined by the present state (three variables) and the control signal (UP). Figure 3.10b shows the required input maps, which have been drawn so that the upper half compares directly with the maps of figure 3.9b. Detailed circuit implementation is left as an exercise for the reader.

3.3.4 Shift registers

This useful 'building-block' circuit configuration is conceptually simple. A suitable number of flip–flops is directly cascaded as shown in figure 3.11. The

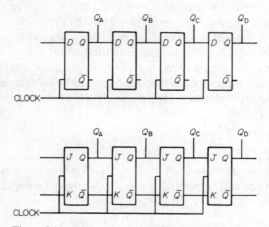

Figure 3.11   *Shift registers using D and JK flip–flops.*

basic property of the shift register is that, after a clock pulse, the $(n+1)$th flip–flop takes on the previous value of the nth, for all flip–flops except the first. The contents of the last flip–flop will be lost (unless transferred to some other storage device) and the new contents of the first flip–flop will be determined by the signal applied to its D input (or to its J and K inputs). In the basic form of the shift register this is the only option available to the system designer. The several system applications, therefore, depend on the origin of these inputs; the possibilities may be categorised and will be discussed in turn.

(*a*) *Constant inputs*   The flip–flops comprising the register will progressively take on, under the control of the clock pulses, the value corresponding to the constant inputs (*ones* for $D=1$ or $JK=10$, *zeros* for $D=0$ or $JK=01$). Notice that, if JK flip–flops are used, there are two more possibilities. $JK=00$ causes the register to fill with the initial value of the first flip–flop whether this be a *zero* or a *one*. $JK=11$ causes the first flip–flop to change state on each clock pulse; hence the register takes on the value 101010 and so on. Although of no great practical application, this last effect often causes confusion when JK inputs have inadvertantly been left open circuit (under which conditions many devices assume the input value *one*). The main application of a constant input to a shift register is to leave the register in the 'all-*zero*' state after a serial transfer of data.

(*b*) *The content of the last flip–flop* To achieve this, the D input of the first flip–flop must be connected to the Q output of the last one (or the J and K inputs of the first connected to the Q and $\overline{Q}$ outputs of the last, respectively). The first flip–flop now takes on the value of the last and the register has become a complete ring. Any data initially loaded into the register will therefore recirculate indefinitely. This property has many applications. For example a stored constant may be used repeatedly in arithmetic processing such as serial multiplication. If one flip–flop is loaded with *one* and all the others with *zero* a ring counter will result having a sequence of states as shown in figure 3.12. This can be used as a simple form of counter but is also useful as a multiphase clock-pulse generator.

Q_A	Q_B	Q_C	Q_D
→1	0	0	0
0	1	0	0
0	0	1	0
0	0	0	1
(1	0	0	0)

Figure 3.12 *Sequence of states for a four-bit ring counter.*

(*c*) *The inverse of the content of the last flip–flop* This property is easily realised by connecting the D input of the first flip–flop to the $\overline{Q}$ output of the last (or the J and K inputs of the first to the $\overline{Q}$ and Q outputs, respectively, of the last). Sometimes known as a 'twisted-ring', 'switch-tail' or 'Johnson' counter, this configuration enables easily decoded counters to be implemented, as shown in the following example.

Example 3.3

Show that a fully decoded octal counter (a counter with eight output lines) can be realised using only four flip–flops and eight two-input gates. What happens if the counter is incorrectly initialised?

Solution Figure 3.13 shows the sequences obtainable by applying complemented feedback to a four-bit shift register. The two sequences are independent of each other and between them cover each of the sixteen possible states once only. Johnson counting makes use of the first of these sequences, which has an obvious symmetry. Decoding each of the eight states apparently requires eight four-input gates, but treating the remaining sequence (8 states) as don't cares reduces this requirement to two-input gates. Using Karnaugh-map techniques, it can be shown that the eight states of the first sequence can

Q_A	Q_B	Q_C	Q_D	Q_A	Q_B	Q_C	Q_D
→0	0	0	0	→0	1	0	0
1	0	0	0	1	0	1	0
1	1	0	0	1	1	0	1
1	1	1	0	0	1	1	0
1	1	1	1	1	0	1	1
0	1	1	1	0	1	0	1
0	0	1	1	0	0	1	0
0	0	0	1	1	0	0	1
(0	0	0	0)	(0	1	0	0)

Figure 3.13   *Sequences obtainable with a four-stage 'twisted-ring' counter.*

be detected as $\overline{Q}_A\overline{Q}_D$, $Q_A\overline{Q}_B$, $Q_B\overline{Q}_C$, $Q_C\overline{Q}_D$, Q_AQ_D, $\overline{Q}_AQ_B$, $\overline{Q}_BQ_C$ and $\overline{Q}_CQ_D$. These reduced terms are valid if, and only if, states in the other sequence do not occur.

Incorrect initialisation will occur if any set of values, other than one of those in the first sequence, is entered into the register (perhaps at switch-on). The system would then cycle indefinitely through the unwanted sequence, producing associated spurious outputs. If the flip–flops comprising the register are of JK type, the counter can be made self-correcting by using an additional gate, to give $J_A = \overline{Q}_C\overline{Q}_D$ and $K_A = Q_D$ (as opposed to $J_A = \overline{Q}_D$ and $K_A = Q_D$). Is is left as an exercise to the reader to verify that this modification does not interfere with generation of the required Johnson sequence, but will divert the spurious sequence into the Johnson sequence after a certain number of counts.

(*d*) *A linear function of the outputs of two flip–flops in the register*   A shift register consisting of n cells is, in general, capable of storing any of 2^n binary numbers; that is, it has 2^n possible states. In all the applications considered so far, for a given initial state, it has not been possible to make the register pass through all the remaining $2^n - 1$ states before returning to its initial state. By feeding back a linear function (that is, one involving only the exclusive OR operation) of the states of the last and other suitable chosen flip-flop this property can be almost achieved. 'Almost' because the state 0000 ..., once entered, cannot be left; the remaining $2^n - 1$ states are entered in an essentially random order. The circuit, therefore, is of use as a pseudo-random-sequence generator. The sequence is described as pseudo-random because, after an interval dependent on the length of the register, the sequence repeats itself (unlike truly random numbers!). A simple example of sequence generation is considered in example 3.4 but a detailed treatment is beyond the scope of this book. An excellent treatment has been produced by Golomb (1967) to which the reader is referred for more details; the more difficult problem of non-linear-feedback shift registers is also considered in this work.

Example 3.4

Determine how a maximum-length sequence can be generated by applying linear feedback to a four-bit shift register.

Solution A two-input exclusive OR gate will be required; one input must be connected to the last flip–flop of the register (otherwise its effective length would be reduced). The sequences obtainable by connecting the other input to the second and third flip–flops are shown in figures 3.14a and b respectively.

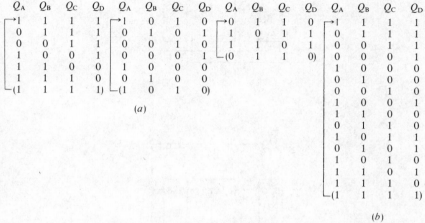

(a)

(b)

Figure 3.14 *Sequences relating to example* 3.4.

It can be seen that connection to the second flip–flop produces three independent sequences, two of length six and one of length three (or four sequences if the state 0000 is regarded as a degenerate sequence of length one). Connection to the third flip–flop produces the required solution, a sequence of length fifteen. It is left as an exercise to the reader to show that connection to the first flip–flop also produces a maximum-length sequence. There appears to be no systematic method of predicting which bit should be used for an n-bit register. Trial-and-error procedures are usually employed.

(e) *An arithmetic function* As mentioned in section 3.3.4b, an arithmetic constant may be recycled and re-used indefinitely. More generally, the contents of the register may be progressively replaced by an arithmetic function of the previous contents of the register and a number stored in another register. Very widely used in the early days of digital computing, this technique is becoming of less importance as digital processors become more and more parallel in operation due to the availability of relatively low-priced medium- and large-scale parallel integrated arithmetic processors.

It is evident from several of the applications that an ability to load data into the register, before any shifting operations, is important. There are many variations on the ways in which this can be achieved; but two distinct approaches can be identified. In one, all cells of the register are first set to the same value, usually *zero*; appropriate inputs are then applied only to those cells that are required to contain a value different from the initial one. A realisation of this approach is shown in figure 3.15a. A pulse on the RESET

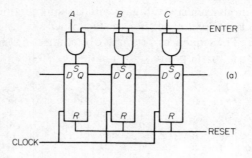

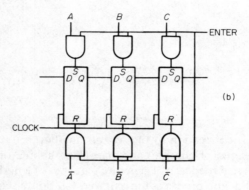

Figure 3.15   *Two methods of entering data into a shift register.*

line sets all flip–flops to zero; subsequently, a pulse on ENTER allows any of the inputs A, B, C and D that are *one* to set the appropriate flip–flops.

The alternative approach is illustrated in figure 3.15b. Here, a single ENTER pulse is required during which any of A, B and C that are *one* will provide set inputs to the appropriate flip–flops whereas those that are *zero* will simultaneously provide appropriate reset inputs and each flip–flop will be given the correct value regardless of its previous one. The latter approach requires more gates but allows data to be entered in a single operation. Note also that in both cases, although the shifting operation is a synchronous one, the data is entered in an asynchronous manner.

3.3.5   *Latches*

Latches are temporary data-storage circuits similar in concept to shift registers but without the facility for shifting data laterally. Consequently, whereas shift registers may in general have either serial or parallel inputs and outputs, latches are restricted to both parallel-input and parallel-output operation. Parallel data may be entered using the techniques described for shift registers.

3.4   Intuitive design of sequential systems

3.4.1   Introduction to design

Well-established formal procedures are available for the design of both asynchronous and synchronous systems and are described in chapters 4 and 5, respectively. In some straightforward cases, a design specification can be met directly by means of an interconnection of counters, shift registers and the like. The method by which such circuits are produced is not easy to define; the required ingredients are familiarity with existing systems of a similar but not identical nature, and imagination. Two examples should illustrate the process.

Example 3.5

Design a synchronous sequential version of the binary comparator outlined in section 3.1.2; the numbers M and N each contain four bits.

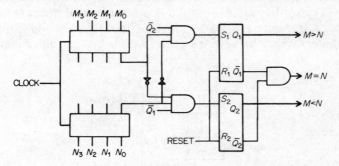

Figure 3.16   *Sequential binary comparator* (*example* 3.5).

Solution   One possible approach is outlined in figure 3.16. Assuming that the data exists in parallel form, shift registers will be required to perform a serial conversion. The most significant digits must be examined first; if they are unequal, the relative magnitude of the two numbers is immediately determined. Therefore $M_0\bar{N}_0$ may be used to set a flip–flop the output of which indicates that M exceeds N; similarly $\bar{M}_0N_0$ can be used to set a second flip–flop the output of which indicates that N exceeds M. If the most significant digits are equal, neither flip–flop will be set and, after a clock pulse, the most significant digits at the end of the register will be replaced by the next most significant. A further comparison will be made, which will result in the appropriate flip–flop being set if one number exceeds the other. Once inequality has been detected, as indicated by the setting of the appropriate flip–flop, it must be ensured that the other flip–flop cannot be influenced by inequalities in less significant pairs of digits. This can be achieved by means of third inputs to the AND gates that inhibit operation once the other flip–flop has been set.

If the two numbers are precisely equal, neither flip–flop will be set and the

AND gate connected to the inverse outputs of the two flip–flops will produce an appropriate output. A practical realisation might need further refinements such as a circuit to ensure resetting, data-entry at the required time and the generation of four, and only four, clock pulses during each comparison.

Example 3.6

Design a sub-system that will divide the repetition rate of an input pulse train by N, where N is a four-bit binary number.

Solution   A counting circuit could readily be designed, using the techniques of example 3.1, to divide the pulse rate by a single, fixed, value of N. A more complex arrangement is required if N can take any value. One approach makes use of a binary comparator, of the type discussed in section 3.1.2, to compare the output of a binary counter with the present value of N. When equality is detected, the counter is reset. If the binary counter is driven by the input pulse train, one reset pulse is produced for every N input pulses; so the reset pulse is also the required output. A possible arrangement is shown in figure 3.17. A difficulty can arise with this circuit; if some flip–flops in the binary counter respond to the reset signal more quickly than others, equality—

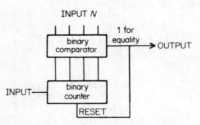

Figure 3.17   *One solution to the pulse rate division problem (example 3.6).*

and hence the reset signal itself—will be lost before the remaining flip–flops are reset. The solution is to use the equality signal either to set a further flip–flop (which can be reset by the next input pulse) or to trigger a monostable circuit (which will automatically reset after a chosen delay). In both cases the effect is to extend the duration of the reset signal so as to allow sufficient time for all flip–flops in the counter to respond.

The cost of the binary comparator can be avoided by using a counter that operates in the reverse mode. The number N is loaded into the counter (using one of the techniques discussed in relation to shift registers in section 3.3.4), which then counts downwards reaching zero after N input pulses. Zero can be detected by means of a single OR gate (four inputs in this case), which is considerably cheaper than a binary comparator (but the reverse counter could be more expensive than a forward one). Alternatively, an up counter could be used in this mode by loading it with the complement of N and counting upwards to zero.

The intuitive approach (illustrated in examples 3.5 and 3.6) generally produces a design much more quickly than the more formal methods to be
discussed later. On the other hand, it is all too easy to fail to consider an
important factor. For example, the additional inputs to the AND gates required to prevent subsequent operation of the flip–flops (figure 3.16) could
easily have been overlooked. Also, intuitive designs rarely use a minimum of
devices; this approach often uses more flip–flops than the theoretical minimum.

The mode of operation of an intuitive design is usually clearly evident from
the circuit diagram; if the design was produced mentally, it can be interpreted
mentally. This is a considerable advantage when testing and fault-finding.
Some formal methods can produce economic circuits of great subtlety whose
mode of operation is not readily apparent.

3.5   Examples

This section contains further worked examples, which provide more details
on counter design, together with two unworked ones. An expanded treatment
of this subject is given by Potton (1973).

Example 3.7

Implement the Gray-code counter of example 3.1 using JK, RS and T flip–
flops.

Solution   Consider first the JK realisation. The state-transition table will be
unchanged from figure 3.9a. From figure 3.5b it can be seen that, if $JK = 10$,
the flip–flop output will be set to *one* and, if $JK = 01$, the flip–flop output will be
reset to *zero*. Therefore, for each flip–flop, the J input must take the value *one*
whenever a *zero*-to-*one* transition is required and, conversely, the K input must
be *one* whenever a *one*-to-*zero* transition is required. These transitions can be
obtained by inspection from figure 3.9a; for flip–flop A, $J_A = \overline{Q}_A Q_B \overline{Q}_C$ and
$K_A = Q_A \overline{Q}_B \overline{Q}_C$. Notice that, unlike the D flip–flop, J and K inputs are not required when the state remains unchanged. For flip–flop B, $J_B = \overline{Q}_A \overline{Q}_B Q_C$ and
$K_B = Q_A Q_B Q_C$ and, for flip–flop C, $J_C = \overline{Q}_A \overline{Q}_B \overline{Q}_C + Q_A Q_B \overline{Q}_C$ and $K_C = \overline{Q}_A Q_B Q_C$
$+ Q_A \overline{Q}_B Q_C$.

Further use can be made of the flip–flop properties in order to simplify
these expressions. Consider $J_A = \overline{Q}_A Q_B \overline{Q}_C$, which is used to bring about the
transition from $Q_A Q_B Q_C = 010$ to $Q_A Q_B Q_C = 110$; if the expression for J_A were
reduced to $Q_B \overline{Q}_C$, J_A would become *one* not only for $Q_A Q_B Q_C = 010$ as required, but also for $Q_A Q_B Q_C = 110$. For this latter case no change in the state
of flip–flop A is required and $J_A = 1$ merely serves to retain the flip–flop in the
required state (K_A will be zero). Even if a transition in Q_A were required, this
could only be from *one* to *zero* and, provided K_A were *one*, $J_A = 1$ would not
inhibit this transition (see last row in figure 3.5b). The simplified expression

for J_A is therefore satisfactory. The general rule, of which this example is a specific case, is that when simplifying an expression for J_i all $Q_i = 1$ may be treated as don't cares; similarly all $Q_i = 0$ may be treated as don't cares with respect to K_i. This makes J_i and K_i independent of both Q_i and $\overline{Q}_i$. Application of this rule reduces the flip–flop input expressions for this example to

$$J_A = Q_B\overline{Q}_C \qquad\qquad K_A = \overline{Q}_B\overline{Q}_C$$
$$J_B = \overline{Q}_A Q_C \qquad\qquad K_B = Q_A Q_C$$
$$J_C = \overline{Q}_A\overline{Q}_B + Q_A Q_B \qquad K_C = \overline{Q}_A Q_B + Q_A\overline{Q}_B$$

which may be used to realise a suitable circuit.

If RS flip–flops were to be used, the input expressions would be the same (replacing J by S and K by R) with one important exception. Since $RS = 11$ is not allowed (unlike $JK = 11$, which produces a transition), the rule for don't cares must be modified to 'all $Q_i = 1$ may be treated as don't cares with respect to S_i except those for which $R_i = 1$' and 'all $Q_i = 0$ may be treated as don't cares with respect to R_i except those for which $S_i = 1$'. This change does not affect the input terms for this particular example.

If T flip–flops were to be used, T must be *one* when, and only when, a transition (whether from *zero* to *one* or vice versa) is required. As with the other single-input flip–flop (D type), there can be no don't cares. Inspection of the state-transition table (figure 3.9a) yields

$$T_A = \overline{Q}_A Q_B\overline{Q}_C + Q_A\overline{Q}_B\overline{Q}_C$$
$$T_B = \overline{Q}_A\overline{Q}_B Q_C + Q_A Q_B Q_C$$
$$T_C = \overline{Q}_A\overline{Q}_B\overline{Q}_C + \overline{Q}_A Q_B Q_C + Q_A Q_B\overline{Q}_C + Q_A\overline{Q}_B Q_C$$

which do not simplify.

Example 3.8

Design one decade of a synchronous binary-coded decimal counter; a carry output is required on the nine-to-nought transition in order to drive further decades. Use JK flip–flops and ensure that redundancies are used as fully as possible to achieve an economical circuit.

Solution   The state-transition table, including the required carry output, is given in figure 3.18a. As values from 10 to 15 inclusive are not used in the BCD system there are six don't cares, which will occur in the same cells in all the flip–flop input maps. Don't cares will also arise from the JK input properties as discussed in example 3.7; the location of these will be different for each map. For clarity, the former don't cares are indicated by $+$ and the latter by $\times$, although in minimisation both are treated in exactly the same way. The required maps are given in figure 3.18b. Note that some cells are don't cares for both of the above reasons.

By inspection, $J_A = K_A = 1$. It is also evident that a carry output is required when $Q_A\overline{Q}_B\overline{Q}_C Q_D = 1$ but use of the don't-care states reduces the required

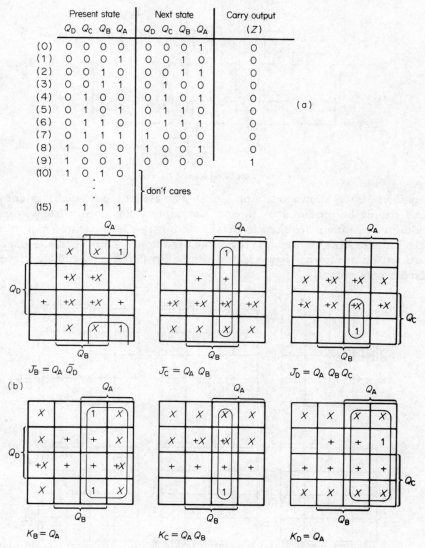

	Present state				Next state				Carry output
	Q_D	Q_C	Q_B	Q_A	Q_D	Q_C	Q_B	Q_A	(Z)
(0)	0	0	0	0	0	0	0	1	0
(1)	0	0	0	1	0	0	1	0	0
(2)	0	0	1	0	0	0	1	1	0
(3)	0	0	1	1	0	1	0	0	0
(4)	0	1	0	0	0	1	0	1	0
(5)	0	1	0	1	0	1	1	0	0
(6)	0	1	1	0	0	1	1	1	0
(7)	0	1	1	1	1	0	0	0	0
(8)	1	0	0	0	1	0	0	1	0
(9)	1	0	0	1	0	0	0	0	1
(10)	1	0	1	0					
⋮									
(15)	1	1	1	1					

(a)

don't cares

(b)

$J_B = Q_A \bar{Q}_D$ $J_C = Q_A Q_B$ $J_D = Q_A Q_B Q_C$

$K_B = Q_A$ $K_C = Q_A Q_B$ $K_D = Q_A$

Figure 3.18   *State table and flip–flop input maps for example 3.8.*

expression to $z = Q_A Q_D$. To ensure synchronous operation of subsequent (more significant) decades, the carry signal would normally be gated with the clock signal. The complete circuit is shown in figure 3.19.

Example 3.9

Design one decade of an asynchronous binary-coded decimal counter. The circuit is to use (unclocked) T flip–flops that change state on the negative-going edge (*one*-to-*zero* transition) of the input pulse.

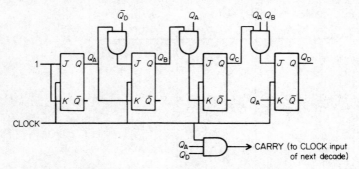

Figure 3.19   *Circuit realisation for example* 3.8.

Solution   It was shown in section 3.3.2 that an asynchronous binary counter can readily be produced by directly cascading T flip–flops. Clearly some additional gating is required to obtain the binary-coded decimal sequence. Using figure 3.18a, the required triggering conditions for flip–flops B, C and D are derived, as shown in figure 3.20a. (Clearly flip–flop A is required to trigger on all input pulses.)

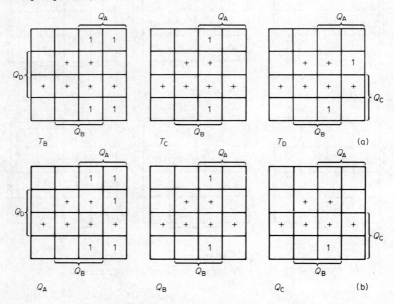

Figure 3.20   (*a*) *Required trigger conditions and* (*b*) *one-to-zero transitions for example* 3.9.

In order to follow, as closely as possible, the simple structure of the asynchronous binary counter (figure 3.7), T_B should be derived from Q_A, T_C from Q_B and T_D from Q_C. To achieve this, the maps of figure 3.20b are constructed; here a *one* indicates where *one*-to-*zero* transitions occur for the flip–flops indicated, when the counter is operating correctly.

Comparison of the two left-hand maps shows that the required trigger conditions for flip–flop B (T_B) can be covered by the *one*-to-*zero* transitions of Q_A but the *one*-to-*zero* transition occurring for $Q_A Q_B Q_C Q_D = 1001$ must be suppressed. This can be achieved by AND gating Q_A with $\bar{Q}_D$; therefore $T_B = Q_A \bar{Q}_D$. The centre maps show that the T conditions for flip–flop C (T_C) can be directly covered by the *one*-to-*zero* transitions of Q_B; therefore $T_C = Q_B$.

The trigger condition for flip–flop D (T_D) is not quite so simple. The condition $Q_A Q_B Q_C Q_D = 1001$ in the map for T_D is not covered by the *one*-to-*zero* transition of Q_C, so an alternative must be sought. Only Q_A covers both conditions required for T_D, so flip–flop D could be triggered from flip–flop A. To eliminate the unwanted transitions of Q_A, $T_D = Q_A Q_D + Q_A Q_B Q_C$ could be used. Alternatively one of the required conditions for T_D could be derived from Q_A and one from Q_C by using $T_D = Q_C + Q_A Q_D$. The carry output is obtained as in example 3.8, but without clock gating of course. A complete circuit is given in figure 3.21.

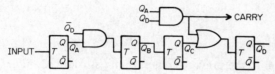

Figure 3.21   *One possible circuit for example 3.9.*

Example 3.10

A memory storage circuit can be formed by cross-connecting a two-input AND gate and a two-input OR gate. Find the truth table for this circuit. Using the inverter-growing technique of chapter 2, show that this prototype circuit can be transformed into either a NAND gate *RS* flip–flop or a NOR gate *RS* flip–flop.

Example 3.11

Develop the synchronous binary-coded decimal counter (example 3.8) to permit reversible operation. Remember that, in the 'count-down' mode, the carry output must be produced on the nought-to-nine transition.

References

Derusso, P. M., Roy, R. J., and Close, C. M., *State Variables for Engineers* (Wiley, New York, 1967).

Golomb, S. W., *Shift Register Sequences* (Holden-Day, San Francisco, 1967).

Potton, A., *An Introduction to Digital Logic* (Macmillan, London and Basingstoke, 1973).

4

Asynchronous Sequential Systems

The discussion in the previous chapter has introduced, in general terms, the concepts involved in both synchronous and asynchronous sequential systems. It has also been shown that many simple systems of this nature can be designed from a purely intuitive point of view. For more complex systems having a larger number of possible operating sequences, the capability of the intuitive approach rapidly becomes exhausted. It is therefore necessary to formulate, as unambiguously as possible, the several steps involved first in analysis and second in design. It is the purpose of this chapter to establish these procedures in respect of asynchronous systems.

The adoption of a formal approach is no excuse for the abandonment of a physical insight into the precise requirements of the systems under consideration. No formal method, however refined, can correct for an initial problem that is specified in an ambiguous or unrealistic manner.

4.1 Basic concepts

4.1.1 General models of asynchronous circuits

In all sequential systems the outputs depend not only on the present values of the inputs but also on their previous values. The variables in the system that 'remember' the past input sequences are the state variables (sometimes called secondary variables). Consequently, at any instant, the past history of the inputs, in so far as this is relevant to future operation, is crystallised in the present value of the state variables.

As explained in section 3.2.1 the property of storage is associated with a feedback path. A general asynchronous system may, therefore, be represented as in figure 4.1. Floating subscripts are used to indicate groups of signals.

Three groups are shown: x_i, the input signals; y_i, the state variables; and z_i, the output signals. Two combinational circuits are shown; one generates the state variables and one generates the output signals.

Since the state variables are binary quantities, the total number of different states that can be represented by q of these variables is 2^q. Let there be p input variables. For every one of the 2^q states, all 2^p different input combinations can, at least in principle, exist. The state of the system, as defined solely by the state variables, is generally referred to as the *internal state*, whereas the state defined by both the state variables and the inputs is referred to as the *total state*. The general model, therefore, has 2^q internal states and 2^{p+q} total states.

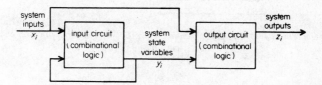

Figure 4.1   *General model of an asynchronous sequential system.*

The sequential operation of the system can be regarded as a series of transitions from one total state to another, under the influence of the inputs. The state variables can be changed only by making an appropriate change of the inputs.

In many cases, for example, a binary counter, the state variables are themselves the required outputs. More generally, the outputs may be associated with each total state of the system. Figure 4.1 shows how the outputs, which are combinational functions of the inputs and state variables, may be generated. There is a theoretical upper limit to the number (r) of distinct output functions that may be generated. The total number of variables feeding the output gating is $p+q$ (p inputs and q state variables). There are, therefore, 2^{p+q} different minterms from which output functions may be constructed; since each minterm may be included in, or excluded from, any one output function, the total number of possible output functions is $2^{2^{p+q}}$. This includes two degeneracies (the logical 'or' of all the minterms $=1$, the logical 'or' of none of the minterms $= 0$). The number of useful functions, therefore, is $2^{2^{p+q}} - 2$. Normally, practical systems would use considerably less than this upper limit.

The model shown in figure 4.1 is quite general; no restrictions are imposed on the feedback connections other than the requirement of a sufficient (non-degenerate) number to store the necessary number of state variables. It is often convenient to localise the storage of the state variables in flip–flops (typically set–reset type). This leads to a general model as shown in figure 4.2, which may be regarded as a special case of the more general model given in figure 4.1.

The input circuit is now required to produce $2q$ signals that form the SET and RESET inputs to the q flip–flops used to store the state variables. Notice that these SET and RESET signals are functions of both the inputs and the state variables.

Localising the storage of the state variables in this manner is convenient from a practical point of view when testing and fault-finding. As will be seen later, it also has the great advantage of completely eliminating static hazards.

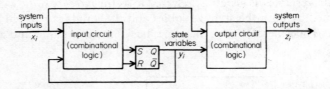

Figure 4.2   *General model of an asynchronous sequential system using flip–flop storage of state variables.*

4.1.2   *Restrictions inherent in the asynchronous approach*

The asynchronous models described in section 4.1.1 are 'self-timed'; that is, changes in state variables consequent on an input change occur as soon after the input change as is permitted by the delays inherent in the input circuit (and the flip–flops in the model of figure 4.2). (Additional delays may sometimes be included; this does not alter the present argument.)

Consequently some restrictions must be imposed on the manner in which the inputs are allowed to change. A second change of input variables must not be allowed to occur before the state variables have settled to the appropriate value following the first input change. If this were not the case the input gating would 'see' the new input combination at the same time as the old set of state variables; clearly this could lead to an incorrect state transition. Since component tolerances can vary over a wide range, the time between successive input changes must be somewhat longer than the worst-case delay of the longest feedback path in the system in order to allow a safety margin.

Secondly, not more than one input variable may be changed at any one time (the 'fundamental mode of operation'). Since absolute simultaneity cannot be achieved, a change from 00 to 11 must be regarded, if time is divided sufficiently finely, as a sequence 00,01,11 or 00,10,11. It cannot be predicted with confidence which sequence will occur and inputs 01 and 10 may well lead to different next states. Hence the state transition is predictable.

In some problems a multiple change of input variables is impossible for physical reasons (for example, excitation by a moving object of two detectors that are some distance apart). In other cases the simultaneous change may be avoided by some operating restrictions (for example, two switches that may not be operated together). If neither of the above applies, asynchronous techniques should not be used (see chapter 5—synchronous systems).

4.2 Analysis techniques

4.2.1 The concept of stable and unstable states

It was shown in section 4.1.1 that a system having p inputs and q state variables has 2^{p+q} total states. These total states can be divided into two classes—those in which the system can remain indefinitely (assuming that an input change does not occur) and those in which the system can remain only transiently. The former are described as *stable* and the latter as *unstable*.

Figure 4.3 shows the general model of figure 4.1 as modified by breaking the feedback connections and substituting hypothetical inputs $(y_1, y_2, \ldots, y_q)$ that feed the input circuit. The output circuit is not shown since it is not relevant to this part of the discussion.

The model is now reduced to a purely combinational one specified by

$$Y_1 = f_1(x_1, x_2, \ldots, x_p; y_1, y_2, \ldots, y_q)$$
$$Y_2 = f_2(x_1, x_2, \ldots, x_p; y_1, y_2, \ldots, y_q)$$
$$\vdots$$
$$Y_q = f_q(x_1, x_2, \ldots, x_p; y_1, y_2, \ldots, y_q)$$

where f_1 to f_q are the q Boolean functions generated by the input circuit. That is, the values of Y_1 to Y_q can be determined for all possible inputs given by x_1 to x_p and y_1 to y_q.

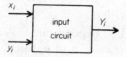

Figure 4.3   *General model of an asynchronous system with the feedback connections broken.*

Of the resulting signals Y_1 to Y_q, two classes may be defined: (1) where Y_i is equal to the postulated value y_i for all values of i from 1 to q. (2) where $Y_i \neq y_i$ for any one (or more) value of i. In class 1 the feedback connection could clearly be replaced without causing any change in the state of the system (this is a hypothetical argument; there may be practical difficulties in remaking the connections without causing disturbances). All states in this class are stable and will exist until a change of input variable occurs.

In class 2 reconnection of the feedback paths would cause at least one of the assumed values of y_i to change to the value of Y_i, which in turn could lead to a change in one, or more, of the state variables; that is, a change of state occurs. States in this class are unstable since they can exist only for a time less than the propagation delay τ around the feedback loop. The circuit keeps changing until a stable state is reached. Instability in many engineering aspects is a disaster to be avoided; in the present interpretation, by contrast, unstable states are essential as the only means of bringing about transitions from one stable state to another.

4.2.2 *The excitation map and flow map*

This section illustrates the concept of stable and unstable states by means of an example and introduces the excitation map as a means of predicting the possible sequences of states. The example chosen (figure 4.4) has two inputs and two state variables ($p=q=2$).

There are, therefore, four internal states and sixteen total states. Derivation of outputs will be considered later.

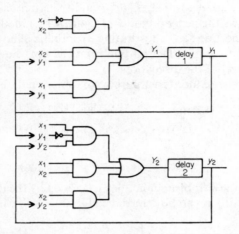

Figure 4.4   *Details of the example discussed in section* 4.2.2.

In the general model, all Y_i depend on all x_i and all y_i; in this particular example, Y_2 depends on x_1, x_2, y_1 and y_2 but Y_1 does not depend explicitly upon y_2. But y_2, of course, still plays an important part in the sequential behaviour of the system.

Boolean expressions for Y_1 and Y_2 in terms of x_1, x_2, y_1 and y_2 may now be written assuming, at this stage, that the feedback paths are not connected

$$Y_1 = \bar{x}_1 x_2 + x_2 y_1 + x_1 y_1$$
$$Y_2 = x_1 \bar{y}_1 y_2 + x_1 x_2 + x_2 y_2$$

The Karnaugh map representing these functions (figure 4.5) is the *excitation map*. For convenience, both Y_1 and Y_2 are indicated on the same diagram. The value of Y_1 is to the left in each cell and the value of Y_2 to the right.

It is conventional to use the columns of the map to represent the inputs x_i and the rows to represent the state variables y_i. Therefore, a change of input implies a change of column while a consequential change of internal state implies a change of row. Each cell of the map represents a total state. Any attempt to mix inputs and state variables in rows or columns leads to confusion because changes in input or internal state appear as diagonal moves across the map.

The stable states in figure 4.5 will now be identified. On the top row

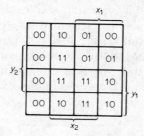

Figure 4.5 *Excitation map where each cell contains the values of $Y_1 Y_2$.*

$y_1 y_2 = 00$ so all cells containing 00 (that is, $Y_1 Y_2 = 00$) will represent stable total states. Similarly on the second row $y_1 = 0$ and $y_2 = 1$ so $Y_1 Y_2 = 01$ denotes stability; proceeding in this manner all the stable states may be identified. These are shown, numbered arbitrarily, in figure 4.6. The circle round the number is a widely used convention indicating that the state is a stable one.

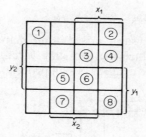

Figure 4.6 *Map showing the stable states.*

The remaining total states in figure 4.6 are, inevitably, unstable. The system cannot remain in any of these states and must respond by moving within the same column until a stable state is located. Transitions between columns are not permissible because the input variables are not allowed to change until a stable state has been reached. The direction of motion within the column is determined by the unstable variables. For example, the top cell of the second column of figure 4.5 indicates instability with respect to y_1. The system will respond in such a manner as to establish equality between y_1 and Y_1; if this can be achieved without changing Y_2 (as is normally the case in good design) there is no reason why a change in Y_2 should occur. Consequently the system responds by moving to stable state 7. This is indicated on the map by entering 7 without a circle (unstable state 7) in the top cell of the second column. Similarly the top cell of the third column indicates instability with respect to y_2, leading to a transition to stable state 3. Proceeding in a similar manner the map is completed as shown in figure 4.7.

This map (the *flow map*) now summarises the entire sequential behaviour of the system. For example, if the inputs are zero ($x_1 x_2 = 00$) the system must be in stable state 1 since this is the only stable state for this input condition. If x_2 is now changed to *one* the system will respond by moving to stable state 7 ($y_1 y_2 = 10$) via unstable state 7. If x_1 is now also changed to *one* the system will move to stable state 6 ($y_1 y_2 = 11$) via unstable state 6. In this manner the sequence of states for a given sequence of inputs can be determined. Remember that double changes of input variable are not allowed.

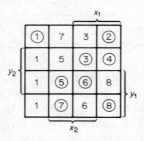

Figure 4.7   *Complete flow map for the example.*

4.2.3   *Flip–flop input maps*

The sequential behaviour summarised in the map of figure 4.7 may be realised using a circuit in which the two state variables are stored in set–reset flip–flops (figure 4.8). The detailed design procedure is given in section 4.4.7.

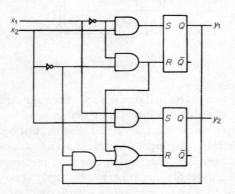

Figure 4.8   *Example using flip–flop storage.*

By inspection of figure 4.8

$$\text{SET } y_1 = \bar{x}_1 x_2 \quad \text{RESET } y_1 = \bar{x}_1 \bar{x}_2$$
$$\text{SET } y_2 = x_1 x_2 \quad \text{RESET } y_2 = \bar{x}_1 \bar{x}_2 + \bar{x}_2 y_1$$

These functions may be represented in map form (figure 4.9).

In the figure R denotes RESET and S denotes SET. Notice that, when the flip–flop concerned is in the *zero* state, an optional reset signal r can be applied if desired and similarly s can be applied when the flip–flop is in the *one* state.

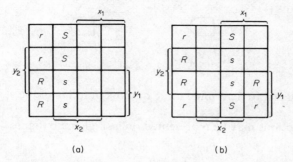

Figure 4.9   *Input maps for (a) the y_1 flip–flop and (b) the y_2 flip–flop.*

This is because these entries have no effect on the flip–flop state and, as will be discussed later, represent don't-care conditions. The capital letters represent a defined change in the state of the appropriate flip–flop. Consequently any cell that contains neither an R nor an S on either of the maps represents a condition where no change is demanded, that is, a stable state. So the stable states may be identified and (arbitrarly) numbered as in figure 4.6. Transitions from the remaining (unstable) states, as shown in figure 4.7, may be predicted using figure 4.9. For example, the total state $x_1x_2y_1y_2 = 0100$ is a SET condition for y_1 and therefore indicates a transition to stable state 7; similarly, the total state $x_1x_2y_1y_2 = 0011$ is a RESET condition for both y_1 and y_2 and so leads to stable state 1. This notation follows that used by Sparkes (1969).

Note that, in the general feedback model, all state variables Y_i depend on all the inputs x_i and all the feedback signals y_i. In the flip–flop model this still applies but the feedback from Y_i to y_i is inherent in the ith flip–flop. In particular cases all dependencies may not be required. In figure 4.4, Y_1 did not depend explicitly on y_2. This results (figure 4.8) in SET and RESET conditions for Y_1 that are independent of both y_1 and y_2.

4.2.4   Output gating and state-transition diagram

The discussion so far has been concerned with the sequential behaviour of the internal states. A simple example of an output circuit will now be examined. Note that, even with only two inputs and two state variables, the total number of distinct non-degenerate output functions is $(2^{2^4} - 2) = 65\,534$ (section 4.1.1). This number includes outputs involving both stable and unstable states. Outputs derived from stable states are levels that remain unchanged until the next state transition; outputs derived from unstable states are brief pulses that exist only for the duration of the unstable state and are not normally useful in practice.

The present example has eight stable states. If only these are used to generate outputs the number of distinct functions becomes $(2^8 - 2) = 254$. By way of illustration two output circuits only will be considered (figure 4.10).

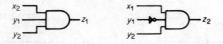

Figure 4.10   *Example of a simple output circuit.*

It is of interest to determine these outputs in terms of the stable states in figure 4.7.

These functions may be represented on a single map (figure 4.11).

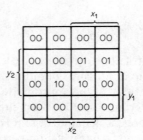

Figure 4.11   *Karnaugh map for the output functions. Each cell represents $z_1 z_2$.*

Comparing figures 4.7 and 4.11, it is evident that output z_1 is *one* in stable states 5 and 6 while output z_2 is *one* in stable states 3 and 4.

The information contained in the flow map (figure 4.7) and the output map (figure 4.11) may usefully be combined in a single diagram known as a *state-transition diagram* (figure 4.12). The notation required for an asynchronous system follows the model described by Moore (1956).

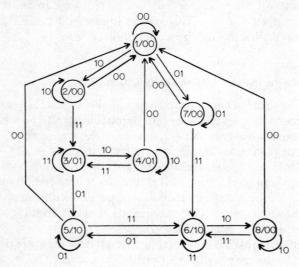

Figure 4.12   *State-transition diagram for the example.*

Each circle represents a stable state. The number in the circle preceding the oblique stroke is the (decimal) number assigned to the state. The binary numbers after the stroke indicate the values of the outputs (z_1 and z_2 in this case) associated with that state. An output derived from an unstable state would be associated not with a particular state but with a state transition as described by Mealy (1955).

The arrows indicate state transitions together with the input value (x_1, x_2) that causes the transition. An essential feature of asynchronous circuits is the arrow returning to the state (usually called a *sling*) indicating the input combination for which that state is stable.

State-transition diagrams will be found useful later as an intermediate stage in converting verbally specified design problems into tabular form, which is required for the subsequent synthesis procedure.

4.3 Races and hazards

4.3.1 Races

The discussion so far has included only a brief mention of the part played by propagation delays in relation to state transitions. Since the timing of asynchronous systems relies solely on these inherent delays, difficulties can arise when two or more signals that have travelled along paths having different delays are combined. In some cases the subsequent state of the system can depend on which of these signals arrives first, that is, which one wins the 'race'.

Races that can lead to a next state differing from that predicted by the idealised theory are described as *critical*. Those that cannot lead to a false state are *non-critical*. Certain specific types of race condition can be identified; these are generally referred to as *hazards* (because of the hazard of maloperation) and have been further categorised as *static*, *dynamic* and *essential*. Each will now be considered in turn in more detail.

Although not mentioned in the original discussion, the first column of figure 4.5 shows a non-critical race. Total state $x_1x_2y_1y_2 = 0001$ is unstable with respect to y_2 and therefore brings about a direct change to stable state 1 (see figure 4.7). A similar argument applies to total state $x_1x_2y_1y_2 = 0010$. Total state $x_1x_2y_1y_2 = 0011$ is unstable with respect to both y_1 and y_2 and therefore a change of both must occur. Due to differing delay times, one must change marginally before the other, giving a transition to either of states $x_1x_2y_1y_2 = 0010$ or 0001; both of these states lead directly to stable state 1 and so the final state is unambiguous even though the route by which it is reached cannot be predicted without a detailed knowledge of the delays within the system.

Figure 4.13 shows the excitation and flow maps for a system having a critical race (and also several non-critical races).

Starting from stable state 1, if x_2 is changed to *one*, the system moves to a total state that is unstable with respect to y_2 thereby predicting a change to stable state 2. If x_1 is now changed to *one*, the state changes unambiguously to stable state 5; changing x_1 back to *zero* initiates a critical race. The state marked with a question mark in figure 4.13 is unstable with respect to both y_1 and y_2. In the unlikely, if not impossible, case of both state variables changing simultaneously, the system will move directly to stable state 2. If Y_2 changes first, a transition to stable state 3 occurs. Since this is a stable state no further change occurs. Conversely, if Y_1 changes first, a transition to unstable state 2 occurs, followed by a transition to stable state 2. Clearly the final stable state depends on the outcome of the race between Y_1 and Y_2.

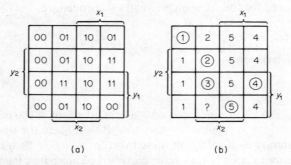

Figure 4.13   *Excitation map (a) and flow map (b) for system having a critical race.*

Such unpredictable behaviour, of course, is undesirable in a practical system. It is not even useful as a generator of random numbers, since a given realisation would have a marked tendency towards one of the stable states. Another realisation using components having different delays (due to production tolerances) might tend towards the other stable state. Good design, therefore, necessitates the avoidance of critical races; further consideration will be given to this later.

Figure 4.13 also demonstrates a further effect that can lead to mal-operation. Although not strictly a race condition its discussion is appropriate here. Consider stable state 5. If x_2 is changed to *zero*, a state that is unstable with respect to y_1 is entered. This causes a change to the top cell of the column, which is unstable with respect to y_2. Stable state 4 is reached ultimately, but after three transitions instead of one. Note also that, during these three transitions, y_1 has changed to *zero* and back again. This effect (a *multiple transition*) is not normally harmful but the needless change of y_1 could lead to trouble if a counting circuit of some kind is fed from this signal.

The possibility also exists of producing an excitation map in which all the cells in one column are unstable. This gives rise (for the appropriate input combination) to continuous oscillation between the various internal states. This is not normally desirable but could be used as a (poor-quality) oscillator.

4.3.2 Static hazards

Although normally discussed in the context of sequential systems the static hazard is purely a property of combinational-logic circuits. It is only when the combinational circuit containing such a hazard forms part of a sequential system that serious mal-operation is likely to occur.

Consider the combinational circuit shown in figure 4.14. By inspection

$$Y = x_1 x_2 + \bar{x}_2 x_3$$

Therefore if $x_1 x_2 x_3 = 111$, $Y = 1$. If x_2 is now changed to *zero*, Y should remain at *one*. This simple argument neglects the extra time delay imposed by the inverter; the consequence of this delay is that the arrival of a *one* (due to $\bar{x}_2$) at the lower AND gate occurs after the removal of the *one* (due to x_2) at the upper AND gate. During this brief period, Y changes to *zero* and back again. The defining characteristic of a static hazard, therefore, is the occurrence of a fleeting change in a variable that, ideally, should not change at all. The delay of a typical inverter could be only a few nanoseconds so the spurious pulse would be quite narrow—difficult to detect, in fact—and unlikely to cause spurious operation of a device actuated by the circuit. In a sequential circuit, due to the presence of 'memory', the spurious pulse can lead to a false state transition and hence can be 'remembered' indefinitely.

If the circuit of figure 4.14 is made into a sequential one by connecting input x_3 as a feedback signal, the excitation equation becomes $Y = x_1 x_2 + \bar{x}_2 y$, giving excitation and state maps as shown in figure 4.15.

Starting in state 5, a change of x_2 from *one* to *zero* should cause a direct transition to state 6. From the excitation map it can be seen that this involves a 'switching-off' of the prime implicant $x_1 x_2$ and a 'switching-on' of the other prime implicant $\bar{x}_2 y$. If this latter term is delayed (in this case by the inverter), the consequent static hazard takes the system to state 3. Since this is

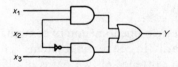

Figure 4.14 *Combinational circuit that demonstrates a static hazard.*

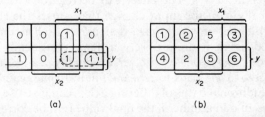

Figure 4.15 *(a) Excitation map and (b) flow map for the system that contains a static hazard.*

a stable state, no further transitions occur and a spurious final state has been reached.

For clarity the static hazard has been explained on the assumption of similar delays for each gate. The unequal propagation times arise because of the presence of an inverter, which introduces extra delay in one path.

Static hazards can also occur in circuits that have apparently equal path lengths for all signals. In this case the differential delays arise from production tolerances of the individual gates comprising the various paths.

The static hazard can be combatted by introducing additional delays (in this case, to the term x_1x_2) using either resistance–capacitance networks or an even number of inverters. More satisfactorily, the effect can be entirely eliminated by the inclusion of extra terms in the combinational logic. The term required in the present example is shown dotted in figure 4.15a; it is a redundant prime implicant from the minimisation point of view but serves, in the present example, to 'protect' the transition from stable states 5 to 6 by generating a term that remains at *one* throughout the transition. This extra term, of course, has no other effect on the sequential behaviour of the system. In general, hazard terms should be included in the combinational expression to 'cover' all required transitions between prime implicants. The complete circuit for the hazard-free version is given in figure 4.16.

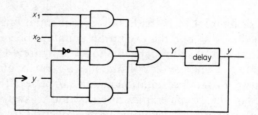

Figure 4.16   *Circuit with additional gate to eliminate the static hazard.*

4.3.3   Dynamic hazards

The dynamic hazard is also a phenomenon of combinational logic that leads to mal-operation primarily in sequential systems. The static hazard arises as a consequence of the existence of two paths of different delay time. The dynamic hazard is an extension of the same concept and involves paths of at least three different delays. The effect can be regarded as a static hazard combined logically with the input change that causes the static hazard. If these occur at different times, the hazard manifests itself as a multiple change in the combined signal. More precisely, a signal that ideally should change once (that is, from *zero* to *one* or vice versa) actually changes three times (for example, from *zero* to *one*, back to *zero* and finally to *one* again). Higher-order effects, due to additional paths of differing delay, can give rise to five or more changes (always an odd number if the final value is to be correct). As with the static hazard, the effect is not normally troublesome in a purely combinational circuit but can lead to mal-operation of an associated sequential system.

Unlike the static hazard, no universal means of elimination exists for dynamic hazards. Since dynamic hazards are the result of multiple path lengths, which themselves arise from complex factoring procedures, it is good practice to use the least possible number of gate delays per path for all paths and to avoid complex factoring. In view of the part played by static hazards in generating dynamic hazards, elimination of static hazards in the first place is clearly helpful in avoiding dynamic ones. In fact it can be shown that, for two-gate-delay logic at least, elimination of all static hazards also avoids dynamic hazards. A thorough treatment has been given by McCluskey (1965).

4.3.4   Essential hazards

Unlike static and dynamic hazards, the essential hazard is associated solely with sequential systems. As a further contrast, the essential hazard is an inevitable consequence of certain specified sequences of states and cannot be avoided by changes in logic design.

The flow-map sequences characteristic of essential hazards are shown in figure 4.17. For these patterns to occur, a minimum of one input and two state variables is required. Of course, similar patterns can arise for more complex systems, in which case the patterns are embedded in larger maps.

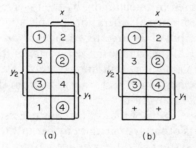

Figure 4.17   *Flow-map sequences characteristic of essential hazards.*

The characteristic feature is that if, starting in state 1, one change of input variable is made, state 2 is reached. If two further changes of input variable are made (that is, a total of three changes), a different state, namely state 4, is reached. During hazardous operation of the system a single input change causes an immediate transition from state 1 to state 4. The effect is primarily a race between the input change and the state-variable changes as will be seen shortly. Clearly the above characteristic pattern cannot be obtained with less than two state variables.

Note that figure 4.17a shows an essential hazard with respect to all stable states whereas figure 4.17b shows an essential hazard only with respect to state 1. In general it is necessary to test all stable states with respect to all input variables in locating essential hazards.

The master–slave flip–flop (figure 4.18) is a widely used circuit that demonstrates an essential hazard. Analysis (which will be left as an exercise to the reader; the required method is indicated in section 4.4.8) shows the flow-map to be as shown in figure 4.17a.

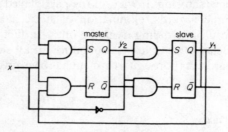

Figure 4.18   *Master–slave flip–flop.*

Assume that the system starts in state 1 and x then changes to *one*. Because of the crossed-over feedback connection, the *one* is applied to the SET input of the master flip–flop causing a change of y_2 to *one*; that is, state 2 is reached. Since $x = 1$, the second pair of AND gates is blocked and no change in the slave flip–flop occurs. When x reverts to *zero*, its inverse permits the slave flip–flop to take the value *one*, following the master, that is, state 3. Finally when x becomes *one* again, the master flip–flop is reset since y_1 is now *one*, and the slave is prevented from changing by the inverse of x.

The above-described normal operation of the system is based on the assumption that the output of the inverter changes *before* the master flip–flop, following a change in x. If the change in the inverter output should follow the change in y_2, the new value of y_2 will immediately be assumed by the slave flip–flop. This corresponds to a direct transition from state 1 to state 4 and is not the required state transition.

To ensure correct operation of the system, it is necessary to include delays in the appropriate parts of the system, thereby permitting the required signal to win the race. If the system is to be produced as an integrated circuit, it is possible to obtain the required relative propagation times by appropriate doping of the various areas of the chip during manufacture.

4.4   System design

4.4.1   *Introduction to synthesis*

So far, analysis of asynchronous systems has been considered together with a discussion of the conditions under which spurious operation may occur. Design techniques based on the methods of analysis will now be developed ensuring (as far as possible) that all spurious modes of operation are avoided.

The first, and often most difficult, part of the design process is to establish precisely the problem to be solved. This may sound rather trite but, particularly in large systems, it is very easy to overlook a particular, perhaps uncommon, operating sequence until very late in the design process or even until testing a completed system.

Of course the general nature of the problem to be solved is known from the first stages but the precise specification must be kept under constant scrutiny. The manner in which the problem is specified can vary widely. Often the problem is described verbally as a sequence of required operations. When the asynchronous system is to be part of an electronic signal-processing system, the specification could be a set of possible input waveforms together with the required outputs (a *timing diagram*).

In all cases the requirement is to obtain a flow table that has the *minimum* number of rows and indicates all state transitions and outputs required to meet the specified problem. Row minimisation will minimise the number of internal variables, and so will usually minimise the amount of logic required. In straightforward cases this table can be written down directly, but often construction of a state-transition diagram is a useful intermediate step. Once the state-transition diagram or flow table has been constructed the design process of row minimisation becomes a relatively automatic procedure. The first step depends on a well-defined problem and, to some extent, the intuition and experience of the designer.

The methods to be described become excessively unwieldy when applied to very large systems. In this case, a flow chart may be required to specify all required operations (see chapter 5 for further details). Partitioning of the total system using this, or some other, approach can usually produce a 'sub-system' specification having a manageable number of inputs and states. The following procedures are suitable for such sub-systems.

4.4.2 *Construction of the primitive flow table*

An example will now be given of the derivation of a state-transition diagram, and hence flow table (or map), from a verbal problem specification. It is required to count the number of people entering or leaving an exhibition. Two light beams are to be used, interruption of each of which generates a *one*. The beams are sufficiently close together for it to be assumed that during entry or exit a normal person will, at some time, interrupt both beams. Two outputs are required, (to drive electromechanical counters)—one for 'entries' and one for 'exits'.

The sequence of interruption will clearly define the direction of motion; the two sequences are 00,01,11,10,00 and 00,10,11,01,00. A trivial solution is immediately apparent since it is necessary only to distinguish 01 or 10 following 00 to determine which of the two sequences is involved. Such an approach would lead to a highly unreliable system since spurious interruption of either beam would immediately generate a spurious output. A more 'secure' solution would be obtained by requiring all three 'non-zero' combinations to occur, in one of the permitted sequences, before an output is gener-

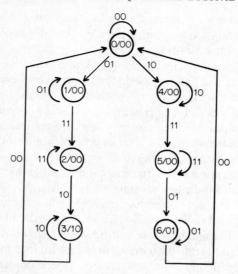

Figure 4.19   *Preliminary state-transition diagram for the example* (section 4.4.2).

ated. A state-transition diagram meeting this specification is given in figure 4.19, where the notation of section 4.2.4 is retained. This diagram is not complete since allowance has been made for the two required sequences but not for any spurious inputs. A successful design can only be produced if, at this stage, proper allowance is made for all conceivable spurious input sequences. It is reasonable to assume in a problem of this kind that a simultaneous change of both detectors will not occur but spurious movements could give rise to several other variable changes. For example, someone could interrupt the first detector and then retreat, giving a return path from state 4 to state 0. A complete state diagram allowing for all possible single-input changes is given in figure 4.20.

The information contained in figure 4.20 may now be transferred directly to a primitive flow table, figure 4.21.

The flow table is said to be *primitive*, since it contains only one stable state per row. All flow tables constructed directly from the problem specification are primitive in this sense; later it will be shown how such tables can be reduced by including more than one stable state in each row.

In many cases the verbal problem specification is related to a waveform, or timing, diagram for the required system. For example, figure 4.22 shows the waveforms relating to a certain pulse synchroniser. An output pulse z is required coincident with the first 'reference' pulse x_1 after the occurrence of an input pulse x_2. It is assumed that x_2 pulses, when they occur, are delayed slightly with respect to the x_1 pulses. In problems of this kind it is important to ensure that the waveform diagram indicates all required operations; the design then proceeds on the strict assumption that these are the only possibilities.

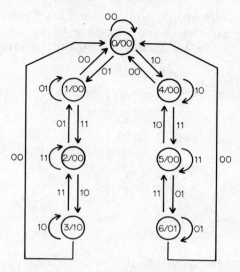

Figure 4.20 *Complete state-transition diagram for the example (section 4.4.2).*

	Inputs ($x_1 x_2$)				Outputs	
	00	01	11	10	z_1	z_2
	⓪	1	–	4	0	0
	0	①	2	–	0	0
	–	1	②	3	0	0
	0	–	2	③	1	0
	0	–	5	④	0	0
	–	6	⑤	4	0	0
	0	⑥	5	–	0	1

Figure 4.21 *Primitive flow table for the example (section 4.4.2).*

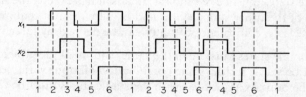

Figure 4.22 *Waveform diagram for the pulse synchroniser.*

By allocating stable states for each input/output combination, as shown in figure 4.22, a primitive flow table may be written down directly (figure 4.23). The seven states shown represent the minimum number required. The 'brute-force' approach would be to allocate seventeen different states and then to apply the techniques to be described in the next section to eliminate redundant states. It is usually possible to eliminate most, if not all, of the redundant states by inspection before constructing the primitive flow table. Care must be taken not to go too far; states 1 and 5 appear to be the same at first sight but state 1 is followed by state 2 and state 5 by state 6 for the same input $x_1 x_2 = 10$

Inputs ($x_1 x_2$)				Output
00	01	11	10	z
①	–	–	2	0
1	–	3	②	0
–	4	③	–	0
5	④	–	–	0
⑤	–	–	6	0
1	–	7	⑥	1
–	4	⑦	–	1

Figure 4.23   *Primitive flow table for the pulse synchroniser.*

and so they cannot be equivalent. If in doubt, allocate an extra state. The reduction procedure will detect any redundant ones.

4.4.3   Equivalent state detection

Having constructed a primitive flow table from the problem data (an ill-defined procedure depending on the nature of the problem itself), the remainder of the design process is relatively automatic. It is always possible to produce an operational circuit, but certain *ad hoc* decisions may have to be taken, which means that a truly minimal implementation cannot always be guaranteed.

The first stage in processing the primitive flow table is to ensure that the number of stable states (that is, the number of rows) is the least required to meet the required problem specification. If this is not the case, one (or more) of the stable states must be redundant; that is, it has exactly the same properties as one of the other stable states and is therefore not required.

More precisely, equivalence of states can be defined by three rules.

(1) The stable states must be in the same column of the table; that is, they must be stable for the same combination of input variables.

(2) All the outputs associated with the stable states must be the same.

(3) All the next states, as specified by the unstable states associated with each allowable input combination, must be the same. Note that, as a consequence of rule 1 above, the allowable input combinations (avoiding multiple change of variable) will be the same for the stable states under consideration.

These three rules can now be applied in turn to all pairs of stable states in the primitive flow table. Rules 1 and 2 are unambiguous, but unfortunately a difficulty can arise with rule 3. Next states that apparently differ may become equal as a consequence of the detection of other equivalent states. Consequently the equivalences that are discovered can depend on the order in which comparisons are made. Furthermore, some of the next states may not be specified (see section 4.4.5); in this case it is acceptable to assume equivalence between a given next state and a don't care. Rule 3 may be modified to allow for these facts.

(3) All the next states, as specified by the unstable states associated with each allowable input combination, must be the same, or equivalent or unspecified.

Figure 4.24 is a primitive flow table that includes examples of all types of equivalent stable states.

Inputs $(x_1 x_2)$				Output
00	01	11	10	z
①	2	–	5	0
1	②	3	–	0
–	10	③	4	1
1	–	3	④	1
1	–	6	⑤	0
–	7	⑥	8⁵	0
1	⑦	9⁶	–	0
1	–	9	⑧	0
–	7	⑨	5	0
12	⑩	11³	–	1
10	⑪	4		1
⑫	14¹⁰	–	13	1
12	–	3	⑬	1
12	⑭	3	–	1

Figure 4.24   *Primitive flow table showing equivalent-state detection.*

State 1 cannot be equivalent to any of states 2 to 11 as a consequence of rule 1. States 1 and 12 satisfy rule 1 but not rule 2 and are therefore not equivalent. Continuing the comparison of all possible pairs of states it will be found that $3 = 11$ since all three rules are satisfied directly. Row 11 is therefore struck out and 11, wherever it occurs as an unstable state, is replaced by 3.

As a consequence of this, 14 becomes equivalent to 10; this equivalence is dependent on the equivalence of 3 and 11. One final difficulty can arise, illustrated by comparison of 5 and 8. These are equivalent except for the differing next states (6 and 9) for input $x_1 x_2 = 11$; comparison of 6 and 9 shows that these states are equivalent provided $5 = 8$. This circular argument is resolved by noting that, provided none of the possible equivalences involved can be shown to be invalid, all may be regarded as self-consistent and therefore allowable. Hence $5 = 8$ and $6 = 9$. The reduced form of the primitive flow table now has ten rows.

It will be realised that, in view of the difficulties associated with the application of rule 3, systematic detection of all redundant states in complex systems is not easy. As an aid, the implication chart, which detects 'implied' equivalences as discussed in the previous paragraph, has been developed by Paull and Unger (1959).

The method involves tabulation of all pairs of states that satisfy equivalence rules 1 and 2. Of these, some will satisfy rule 3 directly giving an immediate equivalence, while others will depend on one or more implied equivalences. Initially all implied equivalences are recorded regardless of their validity.

All implied equivalences are then examined in turn by means of the corresponding cell on the chart. As soon as any one of the implied equivalences is found to be invalid, the possible equivalence of the two states under comparison can be ruled out. As this procedure involves a continuous updating, several 'passes' through the chart may be made before a stage is reached where no further changes can be made. All implied equivalences that survive this procedure can now be regarded as valid. The implication chart for the flow table of figure 4.24 is given in figure 4.25. Note that only half of the square chart is required, since comparison of n with m does not differ from comparison of m with n.

state number	1	2	3	4	5	6	7	8	9	10	11	12	13
2	X												
3	X	X											
4	X	X	X										
5	X	X	X	X									
6	X	X	X	X	X								
7	X	3,9	X	X	X	X							
8	X	X	X	X	6,9	X	X						
9	X	X	X	X	X	5,8	X	X					
10	X	X	X	X	X	X	X	X	X				
11	X	X	√	X	X	X	X	X	X	X			
12	X	X	X	X	X	X	X	X	X	X	X		
13	X	X	X	1,12	X	X	X	X	X	X	X	X	
14	X	X	X	X	X	X	X	X	X	3,11	X	X	X

Figure 4.25   *Implication chart.* √ *denotes unconditional equivalence;* × *denotes non-equivalence on account of rules 1 and/or 2; n,m denotes conditional equivalence.*

4.4.4   *Merging*

Using the techniques outlined in the previous section, a flow table having the minimum required number of stable states can be obtained. The table is still primitive in the sense that the number of rows—that is, required internal states—is equal to the number of stable states. It is the purpose of this section to examine the conditions under which one or more pairs of rows may be combined leading, hopefully, to a reduction in the number of state variables required. Of course the number of state variables required decreases (by one) only when the number of internal states is reduced to less than the next lower power of two (otherwise the number of don't-care combinations is increased).

The process of combining rows in this manner is known as *merging*. Merging is quite distinct from equivalent-state detection and should not be attempted until all redundant states have been eliminated. The problem is to find all pairs of internal states that may be represented by the same combination of state variables without, in any way, changing the sequential behaviour of the system. This is possible if, and only if, the states occurring in each column are the same, either stable or unstable (or don't cares), for the two rows under examination. It follows that two rows satisfying this condition must have their stable states in different columns; otherwise, as all equivalent states have been eliminated, they could not be the same. When merging two rows, stable states take precedence over unstable ones. Note that the outputs need not necessarily be the same.

By way of illustration, the primitive flow table of figure 4.21 will be used. The first and second rows satisfy the criterion for mergeability. On merging, the first row becomes

$$⓪①2\ 4$$

that is, stable and unstable state 0 have combined to give stable state 0. State 1 has been combined similarly and unstable states 2 and 4 have combined with don't cares. Starting in state 0 a change of input to $x_1 x_2 = 01$ causes, exactly as before, a change to state 1. This is now a change of total state but not of internal state, since the change by way of unstable state 1 has been eliminated. Transition to stable states 2 and 4 can still occur as usual by way of the appropriate unstable state. Note that the merged rows inevitably contain entries in all four columns. This does not mean that a direct transition from state 0 to state 2 by way of unstable state 2 is possible; the restriction on multiple-input changes still applies, of course.

Inspection reveals further possible mergers in figure 4.21 and the next stage is to check systematically for all possibilities. Unfortunately, it is not normally possible to make use of all available mergers. For example, the first and fifth rows of figure 4.21 are mergeable but the first row cannot be merged with both the second and the fifth. This is because merging the first two rows assigns unstable state 2 to the don't care in the first row whereas merging the first and fifth rows assigns unstable state 5 to this don't care. Either is permissible but clearly not both. The general rule that can be inferred from the above argument is that if row a is mergeable with row b and also with row c then a, b and c can be merged (a three-way merger) if, and only if, row b is itself mergeable with row c. Furthermore, although rows with differing outputs can be merged, choosing those with the same outputs reduces, or even eliminates, the need for an output circuit.

Selecting the best set of mergers is helped by the use of a merger diagram (see figure 4.26) that shows pictorially all possible mergers. The solid lines indicate mergeable pairs of states having the same outputs (output-consistent pairs) and dotted lines indicate those where the outputs differ. A multiway merger is possible only when all members of the group of states under consideration are mergeable with all others. In the present example, therefore, state 0 is mergeable with 1 or 4 but not both. State 1 is mergeable with both 2 and 3, while 4 is mergeable with both 5 and 6. Note, however, that if these three-way mergers are chosen, then state 0 is no longer mergeable with either 1 or 4.

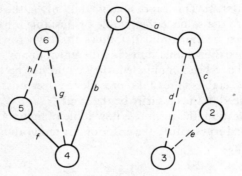

Figure 4.26   *Merger diagram.*

The merger diagram does not provide automatically a best set of mergers; it merely allows the various possibilities to be seen more clearly. In the present example the best solution is to choose the two three-way mergers and leave state 0 unmerged even though this involves making output-inconsistent mergers. If output-inconsistent mergers were not allowed, only two two-way mergers would be possible (for example, 0 and 1, 4 and 5). This would lead to a five-row merged flow table compared with the three-row table of figure 4.27, where the three-way mergers are used. The latter procedure requires one less state variable. If output-inconsistent mergers are used it is not convenient to indicate the outputs on the merged flow table, although some writers do this; see, for example, Unger (1959). The required outputs can still be obtained by reference to the primitive flow table and generated by the output circuit, as will be shown later.

In general, higher-order mergers are possible but a four-way merger is unlikely in a two-input system. This is because a complete row of four stable

	Inputs $(x_1 x_2)$		
00	01	11	10
⓪	1	–	4
0	①	②	③
0	⑥	⑤	④

Figure 4.27   *Merged flow table.*

states would be produced, any of which could be entered via the appropriate unstable state. Exit from this row would be impossible since no input combination could lead to an unstable state that gives rise to an (internal) state transition. Occasional applications of this property can be found—for example, an electronic combination lock, which, once a false combination has been entered, cannot be reset without switching off the power. Normally the effect is to be avoided.

A four-way merger, characterised by mergeability of all pairs of states within the group, would be useful in reducing a system having three or more inputs, and so on for higher-order systems. A four-way merger is recognisable from the merger diagram as a completely connected four-node graph; that is, connections exist between all pairs of the four states concerned.

For large systems, a chart similar to the implication chart may be used to ensure that all pairs of states are checked for mergeability; cell entries would indicate unmergeable pairs, output-consistent mergers or output-inconsistent mergers. The selection of a best set of mergers may also be tackled using a variation of Petrick's method (see section 2.3.6).

Following Petrick it is possible to write a Boolean equation covering all possible combinations of mergers including non-optimal ones. In practice the non-allowed combinations tend to be fewer than the allowed ones and hence easier to treat. In figure 4.26 all possible mergers have been lettered to facilitate reference. Inspection shows that, if a is chosen, b, c and d are not allowable; also if b is chosen, a, f and g are not allowable. In Boolean terms this statement becomes

$$\overline{a(b+c+d)} \cdot \overline{b(a+f+g)}$$

which on simplification using the usual theorems becomes

$$\bar{a}\bar{b} + \bar{a}\bar{f}\bar{g} + \bar{b}\bar{c}\bar{d}$$

This means that either a and b, or a and f and g, or b and c and d must be omitted. Since the first term is least restrictive this would be selected, leaving mergers c to h inclusive as acceptable.

This procedure has automatically included output-inconsistent mergers. If such mergers are not acceptable, at any price, they should be excluded initially. Note that, if there is more than one minimal selection of mergers, this method will indicate them all. It would then be possible to select the combination (if such exists) that involves the minimum number of output-inconsistent mergers.

It is worth recalling that, if a given merger does not reduce the number of internal states required to less than the next lower power of two, no reduction will be made in the number of state variables required; the result is merely an increase in the number of unused state-variable combinations. In this case consideration of the output circuit required may suggest that one or more output-inconsistent mergers should not be made.

4.4.5   Incompletely specified flow tables

The detection of redundant states is further complicated in the case of incompletely specified flow tables. Flow tables for asynchronous systems even when 'completely specified' invariably contain unspecified next states in the columns corresponding to prohibited input changes. In many practical situations, other columns may also contain unspecified (or don't care) next states as a consequence of input sequences that cannot arise in practice. Figure 4.23 developed from figure 4.22 is a typical example of such a situation.

In an incompletely specified flow table the next states that cannot be specified due to input restrictions do not cause any ambiguity. This is because, as a result of rule 1, all equivalent stable states are in the same column as each other, and consequently the unspecified next states are also in the same column as each other. More generally, unspecified next states may occur in any column and may coincide with a specified next state for an equivalent stable state. In declaring these two states equivalent (or compatible), a value is being assigned to what was previously a don't care next state for one of the stable states. This could effectively preclude equivalence between this latter stable state and several other stable states. The situation is analogous to the merger problem where a may be mergeable with either b or c but not both. Now we have $a=b$ or $a=c$ but not both.

The problem is that by choosing, arbitrarily, one pair of equivalences several further equivalences may be ruled out, thereby leading to a nonminimal solution. However, as discussed in section 5.4.1, it is generally better practice to specify a suitable next state in all cases.

4.4.6   State assignment

The processes of equivalent-state detection and merging yield a flow table that has the minimum number of rows consistent with meeting all the requirements of the particular problem specification. It is now necessary to identify each row of the table, that is, each internal state, by means of a unique combination of state variables. For n rows (internal states) q state variables will be required, where $2^q \geqslant n$. If n happens to be exactly a power of two, all possible combinations of the state variables will be just sufficient to identify all the rows of the table. More frequently, the total number of combinations will exceed the number of internal states and several unused or don't care combinations of state variables will arise.

At first sight, assignment of the state variables should not present any difficulty. The internal states could simply be numbered arbitrarily by means of a binary sequence of state variables. Not surprisingly, the problem is not quite as simple as this. Most important, the state variables should be assigned in such a way as to avoid critical races (section 4.3.1); that is, it should never be necessary for more than one state variable to change at once. This will be discussed in more detail later.

Having obtained a race-free assignment, it is not possible to ensure that the combinational logic required (input and output circuits) is minimal for the particular assignment chosen without exhaustively testing all other possible assignments. The magnitude of this task increases very rapidly as the number of state variables increases.

The first assignment may be made in any of 2^q ways, the second in 2^{q-1} ways and so on. Therefore the maximum number of different assignments is $2^q!$ but as there are $(2^q - n)$ redundant combinations this reduces to

$$\frac{2^q!}{(2^q - n)!}$$

Many of these are degenerate since they represent merely an interchange of the labelling or a logical inversion of the state variables, and the number of truly distinct possible assignments has been shown by McCluskey and Unger (1959) to be

$$\frac{(2^q - 1)!}{(2^q - n)!q!}$$

The rapid increase in the number of possible assignments is shown in figure 4.28. Clearly it is impossible to try all the alternatives for any except the very smallest systems. As an aside, note that the last row of figure 4.28 indicates that there are over seventy-five million distinct ways of using four binary digits to represent the ten states of a decimal system.

n	q	$2^q - n$	N
2	1	0	1
3	2	1	3
4	2	0	3
5	3	3	140
6	3	2	420
7	3	1	840
8	3	0	840
9	4	7	10 810 800
10	4	6	75 675 600

Figure 4.28   *Number of possible state assignments* (N); $n =$ *number of internal states*; $q =$ *number of state variables*; $2^q - n =$ *number of don't cares*.

A further consideration that is very important from the practical point of view has received relatively little attention. This concerns the setting up of a conceptual relationship between the state variables and the physical nature of the problem under consideration. For example, a pulse-sorting circuit may be concerned with the order of arrival of pulses on three inputs x_1, x_2 and x_3. From a testing and fault-finding point of view it is particularly convenient if the state variables have an immediate significance such as 'the pulse on x_1 preceded those on x_2 and x_3'. Of course, by the very nature of the state-variable approach any assignment must store all the information of this kind but in many cases the relationships between the state variables and the physical processes can be a very tenuous one. It is the authors' contention that this approach is worthy of serious consideration even if it leads to a non-minimal number of state variables.

Figure 4.29 shows the three possible state assignments for the merged

Inputs $(x_1 x_2)$				Possible state assignments					
00	01	11	10	y_1	y_2	y_1	y_2	y_1	y_2
⓪	1	–	4	0	0	0	0	0	0
0	①	②	③	0	1	0	1	1	1
0	⑥	⑤	④	1	0	1	1	1	0
–	–	–	–	1	1	1	0	0	1
				(a)		(b)		(c)	

Figure 4.29   *Flow table showing possible state assignments.*

flow table developed in section 4.4.4. The total number of assignments is $2^2! = 24$; it is left as an exercise for the reader to establish that the remaining twenty-one of these can be changed to one of those in figure 4.29 by inter-changing y_1 and y_2, logically inverting y_1, logically inverting y_2 or a combination of two or more of these operations.

Assignment b should be avoided since it leads to a race condition. In the case of this assignment, the change from unstable state 4 to stable state 4 requires a change of both state variables; if y_2 should change first, state 3 will be reached and no further change will occur. Similarly assignment c is unsuitable since both state variables change during the transition from unstable state 1 to stable state 1, and so a is the only useful assignment in this case. In general the assignment should be such that all possible transitions from an unstable to a stable state require only one state variable to change. Sometimes this may not be possible, in which case the original design should be carefully examined to see whether any changes can be made to avoid the race condition. Alternatively it may be necessary to introduce an extra state variable; the additional don't cares allow greater flexibility at the assignment stage (see example 4.5).

The bottom row of figure 4.29 consists entirely of don't cares and could, at first sight, be ignored for the remainder of the design. However the system based on assignment a could, just possibly, arrive in the internal state $y_1 y_2 = 11$, possibly on first switching on or due to an interference pulse. (There is, of course, no predicted transition by which this state could be reached.) To ensure consistent operation in the event of such a state being reached, it is helpful to enter unstable states in the row of don't cares in such a manner as to bring the operation back within the required sequence of states. (Stable states could be entered but this would result in the representation of the stable state by more than one combination of state variables.) For example, in the bottom row

0 6 5 4

could be entered giving a reset to stable state 0 in the case of $x_1 x_2 = 00$, and a transition to $y_1 y_2 = 10$ for all other input combinations.

In the present example there is only one race-free assignment and so the design can proceed with the determination of excitation or flip–flop input equations as in section 4.4.7. More generally there will be several distinct race-free assignments and the problem exists of determining the one that leads to the most economical implementation.

In the case of synchronous systems (chapter 5) the restriction on multiple changes of input and state variables does not apply. All possible state assignments are therefore allowable and the problem of choosing an optimum assignment is more difficult. A detailed discussion will therefore be deferred until chapter 5.

4.4.7 Excitation and flip–flop input maps

The design has now reached the stage where a set of state-variable combinations defines the internal states of the system and the remainder of the merged flow table stipulates the required sequences in response to input changes. The associated outputs are defined by the primitive flow table. The remaining design work and subsequent hardware implementation is purely automatic unless the combinational logic involved has more than one minimal solution.

An initial choice must be made between a purely logic-gate implementation (figure 4.1) and one using set–reset flip–flops (figure 4.2); the authors prefer the latter for reasons already stated but for the sake of completeness both approaches will be described.

If a 'gate-only' approach is to be used, the next stage is to construct an excitation map ensuring that the required stable states are, in fact, stable and that the unstable states bring about the required transitions. If state assignment has been done correctly, there will be no critical races but it is useful to make a careful check at this stage.

By way of example, an excitation map for the flow table of figure 4.29 using assignment a will be constructed; note that values (normally unused) will be assigned to the don't cares on the bottom row of the table as mentioned in section 4.4.6. The resulting flow map is shown in figure 4.30, where the lower

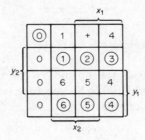

Figure 4.30 *Flow map with assigned state variables.*

two rows of the map have been interchanged in order to give the format required for normal Karnaugh-map representation.

The excitation map can now be constructed; the entries in the cells correspond to Y_1 and Y_2, which will be generated as functions of x_1, x_2, y_1 and y_2. Initially, entries should be made for the stable states (remember that $Y_1 = y_1$ and $Y_2 = y_2$). Figure 4.31 shows the rudimentary excitation map

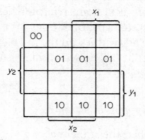

Figure 4.31   *First stage in the construction of the excitation map.*

containing these entries. The remaining entries, apart from the don't care, must be made unstable in such a way as to bring about the required transition. For example, unstable state 1 requires a change to stable state 1—that is, unstable state 1 must be made unstable with respect to y_2—and so 01 is entered. Similarly unstable state 6 requires a change in y_2 (from *one* to *zero*) and so 10 is entered. Note that in practice this procedure is very simple; the binary combination that is entered is precisely that of the stable state to which a transition is required. The resulting complete excitation map is shown in figure 4.32.

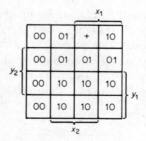

Figure 4.32   *Completed excitation map.*

Combinational expressions for Y_1 and Y_2 can now be written down. To facilitate this the maps for Y_1 and Y_2 can be separated as shown in figure 4.33. From this figure the minimal expressions for Y_1 and Y_2 are

$$Y_1 = x_2 y_1 + x_1 y_1 + x_1 \bar{y}_2$$
$$Y_2 = x_2 \bar{y}_1 + x_1 \bar{y}_1 y_2$$

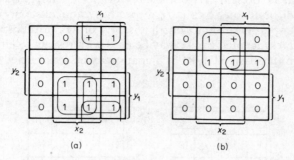

Figure 4.33 *Excitation maps for (a) Y_1 and (b) Y_2.*

where the don't care has been included in both expressions. The system can therefore be implemented as shown in figure 4.34.

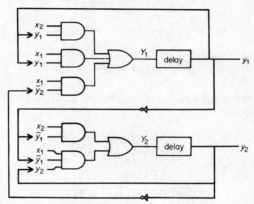

Figure 4.34 *Detailed circuit for the example. The delays represent the total delay of the gates and are not discrete components.*

Alternatively, the system could be implemented using set–reset flip–flops, in which case the flow map of figure 4.30 must be interpreted in such a way as to provide the SET and RESET functions. Stable states, of course, indicate that no change in the state of the flip–flop is required. A system state that is unstable with respect to (normally) one state variable indicates the state variable that is required to change either from *zero* to *one* (SET condition) or from *one* to *zero* (RESET condition). These conditions will be denoted by S and R, respectively.

In addition, don't cares can arise firstly by virtue of unspecified next states as in figure 4.30 and secondly because of redundancy in the flip–flop input equations. That is, once a given flip–flop is in the *one* state it is permissible to apply further SET inputs unless that flip–flop is specifically required to RESET (remember $RS = 11$ is prohibited). In figure 4.30, for example, unstable state 1 represents a SET condition for the y_2 flip–flop but stable states 1, 2 and 3 represent SET don't cares for this flip–flop. SET and RESET don't cares will be denoted by s and r, respectively, and must only be used with respect to the appropriate function, that is, SET or RESET. Unspecified next-state don't cares, denoted by $+$, may be used with respect to either or both, of course. On this basis SET/RESET maps may be constructed as shown in figure 4.35.

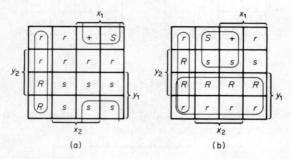

Figure 4.35   *SET/RESET maps for (a) the y_1 flip–flop and (b) the y_2 flip–flop.*

Note that the complementary nature of the flip–flop input equations is such that just one entry arises for each cell (except for $+$, which could also be r or s).

From these maps minimal expressions for the SET and RESET conditions may be derived.

$$\text{SET } y_1 = x_1 \bar{y}_2$$
$$\text{RESET } y_1 = \bar{x}_1 \bar{x}_2$$
$$\text{SET } y_2 = x_2 \bar{y}_1$$
$$\text{RESET } y_2 = y_1 + \bar{x}_1 \bar{x}_2$$

Some writers use a separate map for each function, in which case ones and zeros may be used in place of lettering but twice as many maps are required. From these equations the circuit may be derived directly (figure 4.36).

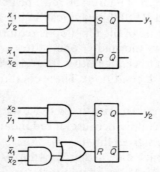

Figure 4.36   *Detailed circuit for the flip–flop version of the example. For clarity, the feedback connections (y_i) and input inverters are not shown.*

4.4.8 Avoidance of hazards

Although care was taken at the state-assignment stage to avoid critical races, no mention was made in the previous section of the care necessary to avoid the various forms of hazard.

Essential hazards, if present in the reduced flow table, cannot be eliminated. At worst, it will be necessary to insert delays at the testing stage to ensure that the appropriate signals always win the race.

Dynamic hazards will be present only if unsuitable (multistage) factoring has been employed or if static hazards are present. It can be seen that the prime obligation of the designer at the stage covered in the previous section is to ensure the avoidance of static hazards.

It will be recalled (section 4.3.2) that static hazards arise when one term, or loop, in a minimised Boolean expression is turning off while another is turning on. The safe procedure is to introduce a further term covering this transition. Inspection of figure 4.33 shows that the present design, by chance, is free from static hazards. In general extra (hazard) terms would be required.

All implementations using set–reset flip–flops are inherently free of static hazards. For example, consider the flow map of figure 4.15b; inspection shows that this could be implemented using a single set–reset flip–flop having

$$\text{SET } y = x_1 x_2$$
$$\text{RESET } y = \bar{x}_1 x_2$$

The corresponding circuit is shown in figure 4.37 where the single set–reset flip–flop has been drawn to show the two NOR gates of which it could be composed.

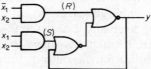

Figure 4.37 *Flip–flop circuit redrawn to facilitate analysis.*

Analysis of this circuit in purely combinational terms yields $Y = x_1 x_2 + x_1 y + \bar{x}_2 y$. Comparison with figure 4.15a shows that all prime implicants are present, including the redundant term required to avoid static hazards. It can be established that set–reset flip–flop implementation automatically involves generation of all the prime implicants, thereby ensuring a design that is entirely free of static hazards. This is a sound reason for adhering to this approach. The extra cost in gates is not large and in this case the flip–flop approach is more economical than the hazard-free feedback circuit (figure 4.16).

4.4.9 The output circuit

The final stage of the design process involves the purely combinational logic that generates the required outputs as functions of the state variables and, in general, the inputs. In some systems, particularly counters, the outputs are the state variables themselves; that is, the output circuit (figures 4.1 and 4.2) has degenerated into a 'straight-through' type of function with no dependence on the inputs.

More generally, a set of output values z_i will be associated with each stable state of the system and, if output-inconsistent mergers have been involved, generation of the output values will require gating of the state variables and input signals. As an example, consider the assigned flow map of figure 4.30; this was obtained by merging the primitive flow table of figure 4.21. Two outputs (z_1 and z_2) are required and, since output-inconsistent mergers have been used, the required values must be obtained by referring back to the primitive flow table. From figure 4.21, it can be seen that output z_1 is required to be *one* only in state 3 and output z_2 is required to be *one* only in state 6. Consequently, from figure 4.30

$$z_1 = x_1 \bar{x}_2 \bar{y}_1 y_2$$

and

$$z_2 = \bar{x}_1 x_2 y_1 \bar{y}_2$$

which can be implemented directly.

In general, a given output may be required to be energised in more than one total state and normal map methods can be used for the minimisation of the Boolean expression involved. Don't care total states may be included but the question of whether or not unstable total states should be included sometimes causes difficulty. For example, the above expression for z_2 could be reduced to $\bar{x}_1 x_2 y_1$ by including unstable state 6 in the output term; since the internal state $y_1 y_2 = 11$ does not normally occur (don't-care assignment), this is perfectly acceptable. Even in situations where the associated unstable state exists frequently, its inclusion in the output expression merely causes the output to change value at the beginning rather than the end of the state transition, an effect that would normally be quite allowable.

Further inspection of figure 4.30 suggests that inclusion of two of the unstable 0 states could further reduce z_2 to $\bar{x}_1 y_1$. The z_2 output would then become *one* as soon as unstable state 6 is entered and remain *one* until the transition from unstable state 0 to stable state 0 is completed. Again this small extension of the output duration would normally be acceptable. Note that this last simplification has introduced a further effect; a change from state 4 to state 0 via unstable state 0 (caused by x_1 changing to *zero*) will now generate an output, momentarily, on z_2 as unstable state 0 is passed through. This may or may not be acceptable, depending on the application. If fast logic elements (for example, transistor–transistor logic) are being used, the spurious pulse will be only a few tens of nanoseconds long and will not affect output devices having a long response time such as electromechanical devices or visual displays (at least those for human perception). If slow logic is being used, and particularly if operation has been deliberately slowed down by the incorporation of time delays, the effect on output devices might be more serious. In any case, if the asynchronous system forms part of a larger system and z_2 could feed further sequential circuits, the spurious pulse could well cause maloperation due to a false state transition.

As a general rule, if there is any doubt concerning the possible effect of including additional unstable states in the output function, the safest practice is to avoid them.

4.5 Examples

This section contains examples, both worked and unworked, based on the material of this chapter. In the worked examples, advantage has been taken of the opportunity to expand some of the points made in the text; the subsequent unworked examples should be capable of solution by readers who have followed the material of the chapter and the worked examples.

Example 4.1

Analyse the circuit shown in figure 4.38 and comment on any abnormal features.

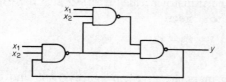

Figure 4.38   *Circuit for example* 4.1.

Solution   On first sight this is an asynchronous sequential circuit having two inputs and one state variable. Assuming the feedback path to be disconnected, the excitation equation (for Y) may be written in terms of x_1, x_2 and y (using the transform rules of section 2.2)

$$Y = x_1 x_2 (\overline{x_1 x_2 y}) + x_1 x_2 y$$
$$= x_1 x_2 (\bar{x}_1 + \bar{x}_2 + \bar{y}) + x_1 x_2 y$$

(by De Morgan's theorem)

$$= x_1 x_2 (\bar{y} + y) = x_1 x_2$$

The feedback variable, here represented by y, is degenerate and cannot be used to store a state variable, and the circuit is not a sequential one at all. Although possibly a 'trick' question, this circuit illustrates a very important point. This is that, although a distinct feedback connection is essential for each state variable to be stored, the apparent existence of a feedback connection does not necessarily guarantee a storage capability. Nevertheless, circuits designed in accordance with the rules presented in this chapter should not contain degenerate feedback loops.

Example 4.2

An asynchronous system is shown in figure 4.39. Analyse this circuit by

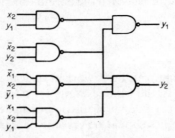

Figure 4.39   *Circuit for example 4.2.*

constructing an excitation map and hence a flow map. This circuit contains two types of hazard. Indicate what these are and explain how mal-operation of the circuit can occur. How would the effect be avoided in a practical circuit?

Solution   Using the transform rules (section 2.2) the excitation equations may be written down directly

$$Y_1 = x_2 y_1 + \bar{x}_2 y_2$$
$$Y_2 = \bar{x}_2 y_2 + \bar{x}_1 x_2 \bar{y}_1 + x_1 x_2 y_1$$

The corresponding maps are given in figures 4.40a and b, respectively. By checking each cell in turn for stability the flow map of figure 4.40c may be constructed. The corresponding state-transition diagram is given in figure 4.40d. (No outputs are shown since they are not specified in the original circuit.)

Figure 4.40 shows that the circuit involves a minimal number of prime implicants, leaving several state transitions unprotected from static hazards. For example, from figure 4.40 it can be seen that the transition from stable state 5 to stable state 8 by way of unstable state 8 involves the 'turning-off' of the term $\bar{x}_2 y_2$ and the 'turning-on' of $x_2 y_1$. If this second term is delayed, Y_1 will 'slip' to zero and remain there, resulting in a transfer to a false stable state (state 4). Inclusion of the hazard term $y_1 y_2$ (shown dotted in figure 4.40a) will protect against all hazards with respect to Y_1. Similarly, the terms $\bar{x}_1 \bar{y}_1 y_2$ and $x_1 y_1 y_2$ are required to protect Y_2 from hazards.

Examination of figure 4.40c shows that there are no critical races but the first two columns show one of the patterns characteristic of an essential hazard. That is, starting in any of stable states 1, 4, 5 or 8, single and triple changes of input variable x_2 result in different final states. For example, starting in state 1, one change of x_2 leads to 4 whereas three changes lead to 8.

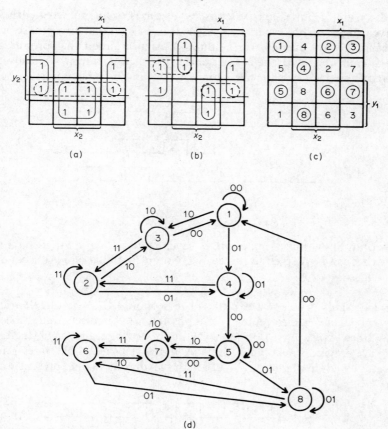

Figure 4.40   *Solution to example 4.2. (a) Map for Y_1. (b) Map for Y_2. (c) Flow map. (d) State-transition diagram.*

This represents correct (predicted) operation; the change from *zero* to *one* of x_2 results in a change from *zero* to *one* in Y_2 and no change in Y_1. Consider the gate (figure 4.39) that has $\bar{x}_2$ and y_2 as inputs; if the signal $\bar{x}_2$ should appear after Y_2 has changed to *one* (due to an excessive delay in the inverter generating $\bar{x}_2$), Y_1 will change to *one* and remain there (due to the term x_2y_1). Therefore, the immediate result is a transition to unstable state 8 followed by an inevitable transition to stable state 8, which represents the final state corresponding to three changes, not one. Correction may be effected by including a delay to ensure that the change in y_2 succeeds that in $\bar{x}_2$.

Example 4.3

An asynchronous sequential system has two inputs x_1 and x_2. Input x_1 is a regular train of pulses and x_2 consists of pulses occurring irregularly and delayed slightly with respect to x_1. The delay is such that an x_2 pulse, when it occurs, always overlaps an x_1 pulse. Design a circuit that produces an output coincident with the next x_1 pulse after the one during which overlap occurs. The details are shown in figure 4.41. Assume no other input conditions

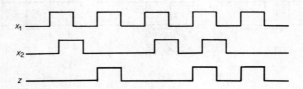

Figure 4.41   *Waveforms relating to example* 4.3.

can occur. Show that a minimum of seven states is required, giving a flow table that can be merged to three rows. Complete the design using set–reset flip–flops.

Solution   This problem was outlined in section 4.4.2. The three variables x_1, x_2 and z have a maximum of eight possible combinations, and inspection shows that two ($\bar{x}_1\bar{x}_2z=001$ and 011) do not occur; at first sight, therefore, there are only six states. The fallacy lies in the fact that stable states 1 and 5 (as numbered in figure 4.42) although having the same present values

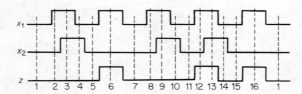

Figure 4.42   *Waveforms of figure* 4.41 *with the states numbered consecutively.*

($x_1x_2z=000$) have different next states (2 and 6) for the same next-input value ($x_1x_2=10$). States 1 and 5 therefore are not equivalent. In deciding, at this stage, that states are equivalent it is important to consider the total sequential behaviour, and not just the instantaneous values of the variables concerned. If in doubt, always assign separate state identifications; the formal equivalent-state detection procedure will identify any equivalences at a later stage. Normally several equivalences can be identified with confidence at this stage but, as an illustration of the process of equivalent-state detection, all combinations of variables in figure 4.42 have been given separate identifications.

Figure 4.43 shows the primitive flow table, which can be written down

Inputs (x_1x_2)				Output
00	01	11	10	z
(1)	–	–	2	0
1	–	3	(2)	0
–	4	(3)	–	0
5	(4)	–	–	0
(5)	–	–	6	0
7	–	–	(6)	1
(7)	–	–	8	0
–	–	9	(8)	0
–	10	(9)	–	0
11	(10)	–	–	0
(11)	–	–	12	0
–	–	13	(12)	1
–	14	(13)	–	1
15	(14)	–	–	0
(15)	–	–	16	0
1	–	–	(16)	1

Figure 4.43 *Primitive flow table derived from figure* 4.42.

directly from the timing diagram. The corresponding implication chart is given in figure 4.44, from which the following equivalences can be inferred: $1 = 7$, $2 = 8$, $3 = 9$, $4 = 10 = 14$, $5 = 11 = 15$ and $6 = 12 = 16$. The reduced flow

State number

	1	2	3	4	5	6	7	8	9	10	11	12	13	14	15
2	X														
3	X	X													
4	X	X	X												
5	2,6	X	X	X											
6	X	X	X	X	X										
7	2,8	X	X	X	6,8	X									
8	X	3,9	X	X	X	X	X								
9	X	X	4,10	X	X	X	X	X							
10	X	X	X	5,11	X	X	X	X	X						
11	2,12	X	X	X	6,12	X	8,12	X	X	X					
12	X	X	X	X	X	✓	X	X	X	X	X				
13	X	X	X	X	X	X	X	X	X	X	X	X			
14	X	X	X	5,15	X	X	X	X	X	11,15	X	X	X		
15	2,16	X	X	X	6,16	X	8,16	X	X	X	12,16	X	X	X	
16	X	X	X	X	X	1,7	X	X	X	X	X	✓	X	X	X

Figure 4.44 *Implication chart for example* 4.3.

table, from which the redundant states have been eliminated, is given in figure 4.45. Note that the unconditional equivalence of states 6 and 12 introduces a 13 into the sixth row of the reduced table. (The reduced flow

Inputs ($x_1 x_2$)				Output
00	01	11	10	z
①	–	–	2	0
–	–	3	②	0
–	4	③	–	0
5	④	–	–	0
⑤	–	–	6	0
1	–	13	⑥	1
–	4	⑬	–	1

Figure 4.45   *Reduced primitive flow table for example* 4.3.

table corresponds to the timing diagram of figure 4.22.) Several mergers are possible; the complete set is shown in figure 4.46, where the solid lines indic-

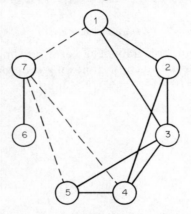

Figure 4.46   *Merger diagram for example* 4.3.

ate output-consistent mergers. For compactness, state 13 has been renumbered 7.

Several combinations of mergers are possible with no clear optimum. Two three-way mergers would be possible only by using (4, 5, 7), which is not output-consistent, and this would lead to a three-row merged table. However, a three-row merged table can be obtained without involving any output inconsistencies. There are two possibilities, namely (1, 2, 3), (4, 5), (6, 7) and (1, 2), (3, 4, 5), (6, 7); the former will be chosen (arbitrarily). The resulting table is given in figure 4.47, where, for reference, the rows have been lettered. The problem now is to assign state variables (two will be required) in a race-free manner. The required transitions may be derived from figure 4.47a and entered on the map of figure 4.47b; this represents a Karnaugh map of the (at-present unassigned) state variables. The asterisk represents the (one) redundant combination of state variables. The transition from a to b represents the transition from unstable state 4 to stable state 4 with similar arrows

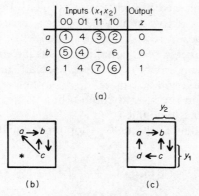

	Inputs $(x_1 x_2)$				Output
	00	01	11	10	z
a	①	4	③	②	0
b	⑤	④	–	6	0
c	1	4	⑦	⑥	1

(a)

(b)   (c)

Figure 4.47   *Merged flow table (a) and inter-row transitions (b) and (c) for example* 4.3.

indicating the other inter-row transitions. Note that, however a, b and c are interchanged on the map of figure 4.47b, at least one diagonal transition remains; this corresponds to a simultaneous change of two state variables (regardless of how they are assigned) and will lead to a critical race.

In this case, the diagonal transition can easily be avoided by bringing into use the don't-care state-variable combination, which will now be identified as row d. This is shown in figure 4.47c, where state variables have been assigned. This assignment is arbitrary but ensures that all required transitions involve a minimum number of state-variable changes. The corresponding modified flow table is shown in figure 4.48; note that unstable state 1 in

	Inputs $(x_1 x_2)$				Output	State variables	
	00	01	11	10	z	y_1	y_2
a	①	4	③	②	0	0	0
b	⑤	④	+	6	0	0	1
c	1	4	⑦	⑥	1	1	1
d	1	4	3	2	+	1	0

Figure 4.48   *Assigned flow table for example* 4.3.

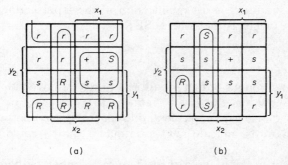

(a)   (b)

Figure 4.49   *Input maps for the flip–flop solution to example* 4.3. *(a) For the y_1 flip–flop. (b) For the y_2 flip–flop.*

row c must be implemented in such a way as to ensure an initial change to row d as shown by the arrow (otherwise a critical race could lead to stable state 5). The remainder of row d has been filled with appropriate unstable states to avoid the possibility of 'jamming' in this internal state. The output for this (purely transient) internal state may normally be regarded as a don't care and, for simplicity, will be equated to y_1 in the present example. (Note that a race can exist in the second column but, as there is only one stable state, this cannot be a critical one.)

From the maps of figure 4.49 the flip–flop input equations are

$$\text{SET } y_1 = x_1 y_2$$
$$\text{RESET } y_1 = \bar{y}_2 + \bar{x}_1 x_2$$
$$\text{SET } y_2 = \bar{x}_1 x_2$$
$$\text{RESET } y_2 = \bar{x}_1 \bar{x}_2 y_1$$

The corresponding system is given in figure 4.50. It is also permissible to

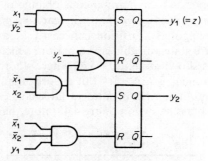

Figure 4.50   *Circuit of the flip–flop solution to example* 4.3.

introduce stable states into the row corresponding to the don't-care state-variable combination. For example, row d could be written as

$$① \; 4 \; ③② $$

In this case, the result is a simplification in the input gating; RESET y_1 becomes $\bar{x}_1 x_2$ (which is the same as SET y_2).

Example 4.4

A certain asynchronous design produces a primitive flow table as shown in figure 4.51. Show that this can be reduced to a merged table having four rows. Derive a logic gate implementation for the system; ensure that this is free of static hazards. Indicate how the outputs (z_1 and z_2) could be obtained.

Solution   By inspection, $1 = 5$ if $2 = 7$, but 2 cannot be equivalent to 7 because of output differences that exclude all other possible equivalences.

Inputs $(x_1 x_2)$				Outputs	
00	01	1↑	10	z_1	z_2
①	2	–	4	0	0
5	②	3	–	0	1
–	2	③	4	1	0
1	–	6	④	1	1
⑤	7	–	4	0	0
–	2	⑥	4	0	1
1	⑦	3	–	1	0

Figure 4.51   *Primitive flow table relating to example* 4.4.

The merger diagram is given in figure 4.52. It can be seen that all mergers are output-inconsistent and the best set is $(1, 4, 6)$ and $(2, 3)$, which gives a merged

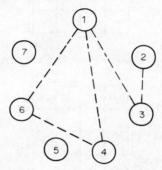

Figure 4.52   *Merger diagram for example* 4.4.

flow table as shown in figure 4.53a. The internal states may be identified as a, b, c and d; the corresponding transitions are indicated in figure 4.53b. At first sight there are two unavoidable diagonal transitions that indicate critical races. However, the transition from c to a occurs in a column containing only one stable state and so the race is not critical. The transition from d to b can be avoided by making use of the don't care on row c. The transition from unstable state 3 to stable state 3 is made to occur by using the don't-care state as shown in figure 4.54 which has the assignment of figure 4.53b. The excitation map (figure 4.55a) may now be written down directly; for clarity, the maps for Y_1 and Y_2 are separated as shown in figures 4.55b and 4.55c. The transitions from state 5 to state 7 and back again involve a static hazard with respect to Y_1; the additional (hazard) term required is shown dotted in figure 4.55b. Similarly, two hazard terms are required in the expression for Y_2 (figure 4.55c). Therefore the full expressions for Y_1 and Y_2 including hazard terms are

$$Y_1 = \bar{x}_1 \bar{x}_2 y_2 + \bar{x}_1 x_2 y_1 + x_2 y_1 \bar{y}_2 + \bar{x}_1 y_1 y_2$$

$$Y_2 = \bar{x}_1 \bar{x}_2 y_2 + \bar{x}_1 x_2 \bar{y}_1 + x_1 x_2 y_2 + x_1 x_2 y_1 + \bar{x}_1 \bar{y}_1 y_2 + x_2 \bar{y}_1 y_2$$

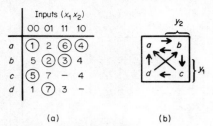

(a)   (b)

Figure 4.53   *(a) Merged flow table and (b) inter-row transitions for example 4.4.*

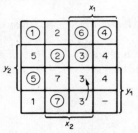

Figure 4.54   *Flow map for example 4.4.*

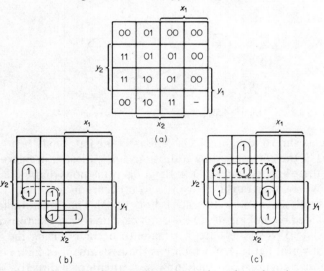

(b)   (c)

Figure 4.55   *Excitation map (a) for example 4.4. Maps (b) and (c) show the required functions for Y_1 and Y_2, respectively.*

Note that the first term is common to both expressions. The outputs (z_1 and z_2) associated with each stable state may be read from figure 4.51 and (using figure 4.54 to locate the stable states) recorded on the output maps (figure 4.56). Note that the don't care associated with the unspecified next state is included, together with don't cares representing the unstable states, which are associated with the appropriate stable states generating the outputs. In this particular example, inclusion of these unstable states as don't cares will not

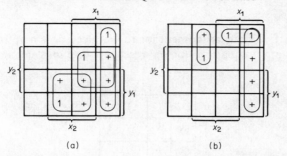

Figure 4.56 *Output maps for example* 4.4. (*a*) *Map representing* z_1. (*b*) *Map representing* z_2.

generate any spurious output pulses, so the minimised output functions are

$$z_1 = x_2 y_1 + x_1 y_2 + x_1 \bar{x}_2$$
$$z_2 = \bar{x}_1 x_2 \bar{y}_1 + x_1 \bar{y}_1 \bar{y}_2 + x_1 \bar{x}_2$$

Note that the term $x_1 \bar{x}_2$ is common to both expressions. By omitting two of the don't cares from the term $x_2 y_1$ in the expression for z_1, the term $x_2 y_1 \bar{y}_2$, which is already available (inverted) as a term in Y_1, may be used. Similarly $x_1 x_2 y_2$ (used for Y_2) can replace $x_1 y_2$. Also $\bar{x}_1 x_2 \bar{y}_1$ is available as a component of Y_2. The resultant system is given in figure 4.57.

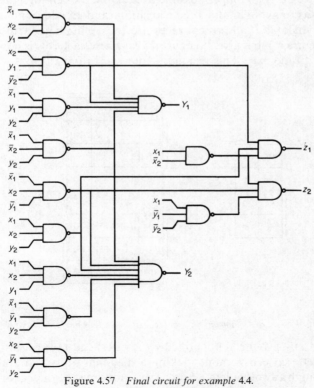

Figure 4.57 *Final circuit for example* 4.4.

Example 4.5

Obtain a race-free state assignment for the reduced flow map of figure 4.58. Hence derive input equations for set–reset flip–flop implementation.

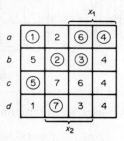

Figure 4.58   *Flow map relating to example 4.5.*

Solution   This flow map is very similar to that derived in the solution of example 4.4 except that all next states are now specified. The required transitions are again as shown in figure 4.53b but the don't care in row c is no longer available and cannot be used to avoid the diagonal transition from row d to row b. It is no longer possible to permute the cells of figure 4.53b in such a way as to avoid all diagonal transitions and an additional state variable is unavoidable if all critical races are to be eliminated. An expanded version of figure 4.53b is given in figure 4.59a, where each row is now assigned to two cells of the map. This technique follows Huffman (1955).

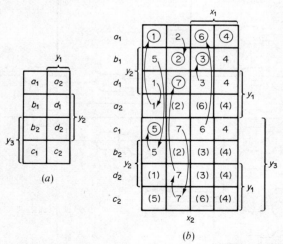

Figure 4.59   *(a) Assignment map and (b) expanded flow map for example 4.5.*

Note that any given cell (identified by a letter) can now be entered from any other lettered cell without making a diagonal transition and without passing through any other letter. Some transitions may be direct, for example,

$b_1 \rightarrow a_1$, but others may involve one or more of the alternative state assignments, for example, $c_1 \rightarrow c_2 \rightarrow d_2 \rightarrow d_1$. Although not essential, it is convenient to associate stable states only with the primary assignments (for example, a_1 includes stable states 1, 6 and 4 together with unstable state 2), whereas the secondary assignment is associated solely with unstable states (for example, a_2 includes unstable states 1, 2, 4 and 6).

Figure 4.59b shows the flow table expanded in accordance with the assignment of figure 4.59a. The rows containing stable states correspond to cells a_1, b_1, c_1 and d_1, which in turn correspond to rows a, b, c and d of the original flow table. The remaining rows, containing unstable states, correspond to cells a_2, b_2, c_2 and d_2 and are essential for the transitions marked by arrows to be free of critical races. Some unstable states on these rows are bracketed; this indicates that they do not take part in any required transition and could be regarded as don't cares. It is more satisfactory to regard them as the appropriate unstable states in order to avoid a possible logic 'jam' if an unfortunate combination of don't cares in the flip–flop input equations should result in extra stable states.

The flip–flop input maps can now be constructed (figure 4.60); care must be taken to ensure that these are constructed in such a way that transitions occur as indicated by the arrows in figure 4.59b, or otherwise critical races will not be avoided.

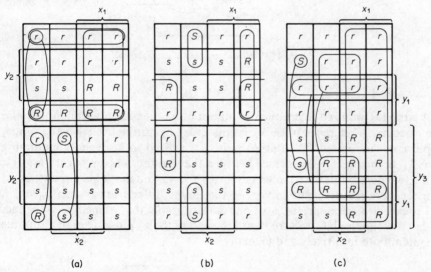

(a)   (b)   (c)

Figure 4.60   *Flip–flop input maps for example 4.5. (a) For y_1 flip–flop. (b) For y_2 flip–flop.*
(c) For y_3 flip–flop.

From these maps the required flip–flop input equations may be written as

$$\text{SET } y_1 = \bar{x}_1 x_2 \bar{y}_2 y_3$$
$$\text{RESET } y_1 = x_1 \bar{y}_3 + \bar{y}_2 \bar{y}_3 + \bar{x}_1 \bar{x}_2 \bar{y}_2$$

$$\text{SET } y_2 = \bar{x}_1 x_2 \bar{y}_1 \bar{y}_3 + \bar{x}_1 x_2 y_1 y_3$$
$$\text{RESET } y_2 = \bar{x}_2 y_1 \bar{y}_3 + x_1 \bar{x}_2 \bar{y}_3 + \bar{x}_1 \bar{x}_2 \bar{y}_1 y_3$$
$$\text{SET } y_3 = \bar{x}_1 \bar{x}_2 \bar{y}_1 y_2$$
$$\text{RESET } y_3 = x_1 + x_2 y_2 + y_1 y_2$$

Example 4.6

Analyse the asynchronous sequential circuit shown in figure 4.61. Express your answer in the form of a flow table showing clearly the stable and unstable states. This circuit contains an 'essential hazard'; explain what this means and show how the existence of the hazard may be recognised from the flow table. How would the effect of the hazard be minimised in a practical circuit?

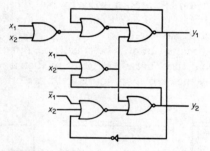

Figure 4.61   *Circuit relating to example* 4.6.

Example 4.7

A circuit is required to produce an output pulse z if and only if an input pulse x_1 occurs. A reference pulse x_2 occurs every operating cycle whether an x_1 pulse occurs or not. The output pulse is required to have the same timing as x_2. Some overlap is assumed between the x_1 pulse, when it occurs, and x_2; the two possibilities are shown in figure 4.62. A combinational circuit cannot meet these requirements in view of the timing difference between x_1 and z. Instead, it is necessary to have one state variable that 'remembers' the fact that an x_1 pulse has occurred until the end of the x_2 pulse. Design a suitable system both intuitively and formally.

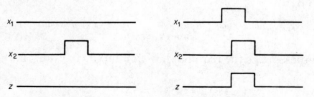

Figure 4.62   *Waveforms relating to example* 4.7.

Example 4.8

Figure 4.63 shows the waveforms relating to a three-bit Gray-code counter (x is the input and y_1, y_2 and y_3 are the state variables). Complete the design using both combinational logic with feedback and set–reset flip–flops. Ensure that static hazards are eliminated by including appropriate terms in the former case. Note that the problem inherently involves essential hazards.

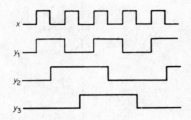

Figure 4.63   *Waveforms relating to example* 4.8.

Example 4.9

A digital phase discriminator has two inputs x_1 and x_2 and two outputs z_1 and z_2. It is required that pulses should occur on output z_1 when the phase of x_1 lags that of x_2; conversely, pulses are required on z_2 when x_1 leads x_2. Suitable waveforms are shown in figure 4.64. Complete the design of the circuit. [An elegant solution using only one state variable has been given by Lind (1972).]

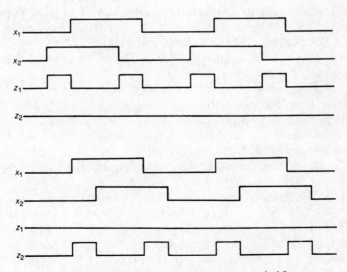

Figure 4.64   *Waveforms relating to example* 4.9.

Example 4.10

The primitive flow table for an asynchronous sequential system is shown in figure 4.65. Eliminate all redundant states. Show that the reduced flow table can then be merged to a table containing three internal states. Note that one of these has no possible exit. Obtain a race-free state assignment and complete the design of the system.

Inputs $(x_1 x_2)$				Output
00	01	11	10	z
①	6	–	11	0
②	6	–	10	0
③	4	–	10	0
3	④	9	–	1
2	⑤	9	–	0
1	⑥	8	–	1
–	5	⑦	11	0
–	6	⑧	11	1
–	4	⑨	10	1
3	–	9	⑩	1
1	–	8	⑪	1

Figure 4.65   *Primitive flow table relating to example* 4.10.

References

Huffman, D. A., *A Study of the Memory Requirements of Sequential Switching Circuits*, Massachusetts Institute of Technology, Technical Report 293 (1955).

Lind, L. F., 'An Improved Digital Phase and Frequency Sensitive Detector', *Proceedings of the I.E.R.E. Conference on Digital Processing of Signals in Communication* (1972) pp. 277–84.

McCluskey, E. J., *Introduction to the Theory of Switching Circuits* (McGraw-Hill, New York, 1965) pp. 299–306.

McCluskey, E. J., and Unger, S. H., 'A Note on the Number of Internal Variable Assignments for Sequential Switching Circuits', *I.R.E. Trans. electronic Comput.*, 8 (1959) pp. 439–40.

Mealy, G. H., 'A Method for Synthesising Sequential Circuits', *Bell Syst. Tech. J.*, 34 (1955) pp. 1045–79.

Moore, E. F., 'Gedanken-experiments on Sequential Machines', in *Automata Studies*, ed. C. E. Shannon and J. McCarthy (Princeton University Press, 1956) p. 134.

Paull, M., and Unger, S. H., 'Minimising the Number of States in Incompletely Specified Sequential Switching Functions', *I.R.E. Trans. electronic Comput.*, 8 (1959) pp. 356–367.

Sparkes, J. J., *Transistor Switching and Sequential Circuits* (Pergamon, Oxford, 1969).

Unger, S. H., 'Hazards and Delays in Asynchronous Sequential Switching Circuits', *I.R.E. Trans. Circuit Theory*, 6 (1959) pp. 12–25.

Further Reading

Friedman, A. D., and Menon, P. R., *Theory and Design of Switching Circuits* (Computer Science Press, 1975).

Lewin, D., *Logical Design of Switching Circuits* (Nelson, London, 1970).

Marcus, M.P., *Switching Circuits for Engineers*, 2nd ed. (Prentice-Hall, Englewood Cliffs, N.J., 1967).

5

Synchronous Sequential Systems

The formal basis and development of combinational-logic design and asynchronous design have been discussed in previous chapters. In this chapter attention is given to the design of clocked or synchronous systems. Extensive use of medium- and large-scale integration in logic systems is encouraged in this design procedure.

The techniques developed for asynchronous systems are also applicable to synchronous design, with a few differences. The first significant difference is that, with synchronous designs, two or more input signals can quite freely change between clock pulses, whereas fundamental mode operation is assumed for asynchronous designs. In the initial design, therefore, there are more permissible input combinations to consider. Another important difference is that for synchronous logic all states are considered to be stable, and so the concept of an unstable state disappears. The possibility of hazards and races is therefore completely eliminated. Other aspects of design—for example, the state-transition diagram, the flow map and detection (and elimination) of equivalent states—remain the same.

5.1 Advantages and disadvantages of synchronous systems

The basic idea behind the synchronous concept is as follows. Imagine a train of narrow positive pulses, which will be referred to as *clock pulses*. These pulses are normally equally spaced from each other in time. A synchronous device will commence activity only at the beginning of (or during) a clock pulse. It will finish its processing at some time after the pulse, but is guaranteed to be finished *before* the arrival of the next clock pulse. When many such devices are connected to a common clock pulse, they will all start operation (or be synchronised) at the arrival of a clock pulse, and will finish their work at more or less random times. But each device will have a worst-case upper time limit, after which time it is guaranteed that it has finished its operation. The time between successive pulses should be arranged to be slightly longer than the worst-case time of the slowest device in the system, to ensure that the system functions properly. Thus the tolerances in the operation times of the various devices have been accounted for.

From an equipment-manufacturing point of view, this state of affairs is very desirable. It is not necessary to measure the actual speed of operation of each device. Also from a maintenance and testing point of view, a synchronous system is highly desirable. If a device should fail, it can be removed and another unmeasured device of the same type number can be put in its place, with the knowledge that the system should again function properly. To test a synchronous system it is sufficient to note that any clock-pulse rate slower than the specified rate will also result in correct system operation. This fact means that clock pulses can be applied, one at a time, by the tester. After each pulse the system will remain static until the next pulse is applied, thus allowing leisurely inspection for fault detection. Another possibility is to use a fairly low pulse rate and observe the outputs on an inexpensive oscilloscope.

If this state of affairs is compared to a system that is entirely asynchronous, the advantages of the synchronous system become apparent. In an asynchronous system the time delays (both propagation delays and processing delays) of the various devices are very important and, in the feedback loops, play a crucial role in ensuring proper system operation (recall the discussion on hazards in section 4.3). If a critical race is discovered in the design, for example, this snag will have to be resolved either by redesign (which will usually result in more gates), by device characterisation (for example, ensure that gate A has less than twice the propagation delay of gate B), or by adding additional components into the system. This latter method consists of adding capacitors in shunt with the gate outputs that are to be slowed down. Then the rise time of the gate will be controlled mainly by a resistance–capacitance time constant, with the capacitance externally controllable.

Needless to say, the above techniques can be time-consuming and costly. It is usually the case that in a rather large asynchronous system the interactions between its various parts can be very complex to analyse. In equipment manufacture one cannot merely state that if the first twenty asynchronous units function correctly then the next unit is bound to work. Sooner or later, a unit will be built with all the time delays at the wrong extremes, in which case the unit might well fail its production test. When this failure occurs, then it is necessary to resort to the modifications mentioned above, all of which are costly to the manufacturer. If a device in the asynchronous system should fail, the replacement device (although within the manufacturer's specification) could lead to system failure, due to delay-time differences. Finally, in testing an asynchronous system it is very difficult (if not impossible) to slow down the circuit action for observation and recording purposes. Therefore, expensive equipment is required for this task.

The outstanding advantage of asynchronous circuitry is the rapid speed with which it operates. To get the best of both worlds, the prevailing design philosophy is to use systems that are synchronous, but circuitry within the system that is asynchronous. In medium-scale integration this philosophy is followed. A range of clocked (or synchronous) medium-scale integration devices is available in the various logic families. Close inspection of the device circuitry reveals that asynchronous techniques have been employed in their design. If the overall system (which includes medium-scale integrated circuits) is synchronous, and a suitable clock rate is specified, there is no limitation to the number of chips that can be used, and hence a very complex system can be developed. Third-generation computers are an example of this development. The one disadvantage of a synchronous system is the relative slowness of operation, which is due to the fact that the time between successive clock pulses must be longer than the longest processing time of the slowest device in the system. Various methods to cope with this problem have been developed. In some logic families there are certain members that have been specifically developed for high-speed operation, such as Schottky-barrier versions of transistor–transistor logic. These devices are pin-compatible with their lower-speed counterparts in the family. Another technique that proves useful in some cases is to redesign the system to include more parallel processing. One can often obtain a speed increase for a cost penalty (more devices) in this manner. Finally there is the possibility of mixing different logic families together. This procedure is not without its disadvantages, notably the interfacing problem and different power-supply requirements. Still, the virtues of each family can be utilised by this means. For example, the high-speed portions could be realised with emitter-coupled logic, and the lower-speed portions (which might involve complicated processing operations) with transistor–transistor logic or, speed permitting, complementary-metal-oxide-semiconductor logic.

5.2 Preliminary design

5.2.1 Initial specification

The specification for a synchronous system is often given in word form. The main problem with a word specification, as mentioned in the previous chapter, is that it does not usually give a complete listing of all input combinations and sequences and their associated outputs. As an example, consider the specification

Design a synchronous system which has two inputs, x_1 and x_2, and an output z; z is to be turned on if and only if x_1 goes high before x_2.

This specification is ambiguous. Consider the input sequence $x_1 x_2 = 00$, 10, 00, 10, 00. When the inputs change from 10 to 00, should z go low, or remain high? There are various possibilities to consider.

(1) The specifier knew that this sequence would not occur in practice and so did not bother to include it in the specification.
(2) The specifier knew that this sequence would occur, but is not concerned with the output behaviour for this case.
(3) The specifier did not consider the consequences of this sequence occurring.

Because of possibility 3, at this stage the designer's best plan is to contact the specifier, if possible, in order to clarify the situation. If the specifier is not available, the designer will have to make some assumptions as to the intended use of the system and proceed accordingly. Ideally speaking, the designer would prefer to work from a state-transition diagram that would list all of the possible combinations. The above specification could be made more definite by either

(1) expansion of the word statement of the problem to include the output value for all possible input patterns

or

(2) the use of a flow map instead of a word statement to define the problem.

The point to be emphasised here is that it is vitally important that the designer be given (or obtain) a clear and *complete* specification. It is not satisfactory to accept a muddled specification and then to carry on with the expectation that somehow things will sort themselves out.

5.2.2 Synchronous-system description

The general model of a synchronous system is given in figure 5.1. This figure should be compared with the asynchronous-system model of figure 4.2. In the synchronous-system model, floating subscripts are again used to indicate groups of signals. For example, x_i represents the group of all input signals. Four groups are so identified: x_i, the input signals; y_i, the system-state signals; z_i, the system-output signals; and w_i, the actions-output signals.

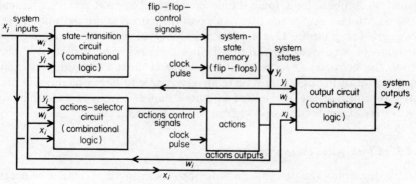

Figure 5.1 *General model of a synchronous system.*

The state-transition circuit consists of combinational logic. This block has inputs x_i, w_i and y_i. Its purpose is to provide the proper input signals for the flip–flops that comprise the system-state memory. As will be seen, this memory is analogous to the instruction steps in a computer program.

The system-state memory is used in conjunction with x_1 and w_i to control the actions block by means of the actions-selector circuit. This selector circuit, which is composed of combinational logic, decodes its input information and drives appropriate actions-control lines. When the next clock pulse arrives, the appropriate actions are initiated.

The provision of an actions block allows the designer to make full use of the medium- and large-scale integration that currently exists. This block can in principle contain circuitry of any complexity. It could contain the action of a small asynchronous system (which is triggered by the clock pulse), or an entire computer processor, for example. When an action is completed, it might be desired to send an 'action-completed' signal back into the system, as a safeguard for proper system operation. The w_i signals are used for this purpose. They can also be used to indicate a defined stage of some activity (for example, a counter reaching a specified count) in the actions block.

The circuit that provides the system-output signals z_i is also shown in figure 5.1. The inputs to this combinational-logic circuit are normally y_i and x_i, although one or more of the w_i signals may occasionally be needed.

As stated above, the system-state-memory block corresponds to the instruction steps of a computer program. The system changes its state from one clock pulse to the next in the following manner. The y_i outputs are applied to the state-transition circuit. This circuit combines the y_i signals (along with x_i and possibly w_i), which are in turn connected to the system-state memory. A complete feedback loop exists (in some ways similar to the feedback loops of asynchronous systems), which is capable of indefinite memory storage. The x_i signals are used to control the sequencing of system-state memory. As an example, suppose that the system is in state s_1 (given by $y_1 y_2 y_3 = 101$, say). On the next clock pulse, if $x = 1$ the system is required to go to state s_2 ($y_1 y_2 y_3 = 110$) or, if $x = 0$, to go to state s_3 ($y_1 y_2 y_3 = 000$). Then the state-transition combinational logic should be designed to meet this requirement.

As mentioned above, the range of possible actions is limited only by the designer's imagination. It is a good plan to check on what is available in medium- or large-scale integration before embarking on a design. If each action can be made to correspond to one, or at most a few, integrated circuits, the resulting design will be relatively simple to develop, and will take full advantage of the specialised circuitry that is currently available. An illustrative example of this procedure will be given shortly.

5.2.3   Clock-pulse cycle

Before describing the design technique for synchronous systems it is necessary to look at the sequence of events from one clock pulse to the next. This sequence is shown in figure 5.2 for master–slave synchronous operation. It is convenient to start the description during time period A_1, for which the output signals of the state-transition circuit and actions-selector circuit (in figure 5.1) are assumed to have settled. These signals can be viewed as giving specific orders or commands to the synchronous devices in the memory and action boxes during time period B_1, which includes the rising clock-pulse edge. The outputs of these devices, however, are not allowed to respond to their orders during this time. These orders are usually stored internally in 'master' storage units, which are composed of flip–flops. The duration of the clock pulse is long enough to ensure that the orders are stored properly. The next event to occur is the falling edge of the clock pulse. This edge has two functions. It blocks the masters from receiving fresh information on the input lines until the next positive-going clock-pulse edge. It also starts a flow of information from the master units to the slave units. The memory y_i goes to the next state. At the same time, the previously stored actions are executed. Then

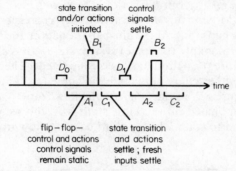

Figure 5.2   *Sequence of events for three typical clock pulses.*

new w_i signals are formed by the actions box. The new w_i, y_i, and new input signals are assumed to settle during C_1. These signals are applied to the state-transition circuit, and to the actions-selector circuit, which are combinational-logic circuits with inherent time delays. During the time period D_1 the outputs of these circuits settle. The above pattern then repeats with A_2, B_2, C_2 and so on.

The important point to note is that the system-state memory flip–flops are constrained. An output change initiated by a clock pulse will produce an immediate input change by means of the state-transition circuit, but does not produce a second output change until the next clock pulse. This is in contrast to an asynchronous system, where the memory variables keep changing until a stable state is reached. The clock pulses therefore act as regulators of flip–flop activity.

5.3   Flow-chart method of design

5.3.1   *Introduction*

From the preceding timing sequence of figure 5.2 and from the synchronous model in figure 5.1 it is possible to construct a chart of synchronous-system activity. Such a chart will henceforth be called a flow chart, as it will depict the flow of system activity from one state to another. The general topology of this chart is shown in figure 5.3. The system states are indicated by double horizontal lines. One can imagine that the system is in one of its states during the time periods D_0 and A_1 on figure 5.2. When a clock pulse occurs, flow commences on the flow chart from state j, which has uniquely specified y_i values. As a general rule, there should be only one flow path leaving any state on the chart. Usually it is convenient to associate downward movement with increasing time. In figure 5.3 the flow from state j encounters a number of binary decisions based on x_1 and w_1, where the w_i values are a result of actions during the previous clock pulse. These decisions are used to direct the flow to one particular action. These decisions are made by the actions-selector circuit during B_1 in figure 5.2. This flow can be considered to occur at the start of time period C_1 in figure 5.2.

Having selected an appropriate action, the flow is next directed to one of the actions boxes. Moving through an actions box corresponds to some of the time interval C_1 in figure 5.2. At the finish of the action, the flow must terminate on another (or possibly the same) system state. (In reality the new state usually occurs before the action has been completed.) The flow then rests at that state until another clock pulse arrives. Three typical flow terminations are depicted in figure 5.3. In the first case the flow from two separate actions (A_1 and A_2) is terminated on the same state k. In the second case the flow from A_3 is directed to state l. In the third case the flow from A_4 is directed back to state j.

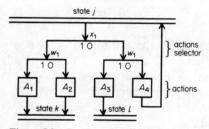

Figure 5.3 *Portion of a typical flow chart.*

In figure 5.1 it will be seen that an actions block is given as a separate entity from the system-state memory, even though the actions block will normally contain flip–flops. If one adopts the classical definition of the state of the system as being determined by *all* flip–flop variables, then the actions block could also contribute to the system state. The following example should make the distinction between the two blocks clear. It is desired to increment a twelve-bit counter, if a specified input exists, on each clock pulse. If the counter is considered to be in the system-state memory block, it will add twelve state variables, or $2^{12} = 4096$ states to the system. The resulting state diagram would be most unwieldy. If the counter is placed in the actions block, however, the above complication is avoided. Now there is only one device to be listed when describing system operation.

At this stage of design the controller that moves the system from one state to another is not considered in great detail. This controller is the state-transition circuit in figure 5.1, and will be considered in more detail when the state-assignment problem is discussed. For the present discussion it is reassuring to note that this combinational-logic circuit can in principle be a read-only memory. For this case the input variables x_i, y_i and w_i specify a memory address. The contents of that address are the control signals to the system-state memory. Standard combinational-logic design techniques can also be employed for this purpose.

The flow-chart technique is a powerful design aid. It displays the decisions and actions of a synchronous system in a clear manner, and also allows the designers to take full advantage of the specialised integrated circuits that now exist. There is no reason to limit the actions boxes to contain only digital circuitry. In principle analogue devices could also be used, provided their actions were completed in time interval C_1. Thus the flow-chart method can also be employed in the design of hybrid analogue/digital systems. The flow chart is also similar to the flow chart that is used in computer programming, a good description of which is given by Knuth (1972). In most cases (depending on the required actions) it will be possible to produce either a computer program or a hard-wired synchronous system from the same flow chart. The relation of the flow chart to a computer program should be clear from the above description.

5.3.2 Phase design

The specification of a synchronous system usually involves a detailed description of the system in various phases of operation. For example, a data-retrieval system could have a data-acquisition phase and a data-transmission phase. A natural way of approaching this sort of problem is to break the system down into a number of phases, design a flow chart for each phase, and then to link these phases with a final flow chart. This approach enables the designer to solve a number of small problems, rather than one large over-all problem. The same sort of procedure is used in computer programming. Various sub-routines are written and debugged. These sub-routines are then linked by a main program which, in some cases, is quite short.

This method will be illustrated with the following example.

Example 5.1

Design a synchronous system that will record and transmit information related to signal x_1. Signal x_1—the output of some physical-parameter-to-frequency transducer—is assumed to have a one-to-one mark-to-space ratio; that is, it is a square wave. A control signal x_2 is also provided. When

$x_2 = 0$, the system is required to record the highest frequency that is reached by x_1. During this time the system output z is to remain at *zero*.

When x_2 goes to *one*, the system is required to produce an output z that is a square wave. The frequency of z is the (highest) recorded frequency of x_1 during the time when $x_2 = 0$.

In this example the phases of operation are evident. The first phase ($x_2 = 0$) will be called the *acquisition phase*, and the second phase ($x_2 = 1$) will be called the *transmission phase*. The acquisition phase will be considered first.

Recording the highest frequency of x_1 is equivalent to recording the shortest period of x_1. As x_1 is a square wave, the shortest half-period could also be recorded. The half-period measurement is convenient, as it allows a measurement to be made during $x_1 = 1$, say, and allows for data updating to be done during $x_1 = 0$. The half-period of x_1 can be conveniently measured by counting pulses, as long as these timing pulses are at a much higher frequency than x_1. The timing pulses themselves can be used as clock pulses for the synchronous system in order to simplify matters.

With these thoughts in mind, a catalogue of medium- and large-scale integrated circuits is consulted. It is found that counters, data registers, and digital-magnitude comparators are readily available, and so can be used in the actions block of the system. The frequency requirements for the clock pulse (which will be a function of x_1 and the required system accuracy) will indicate which logic family to employ.

A preliminary flow chart for the acquisition phase can now be constructed. Refer to figure 5.4 for details of the construction as the chart is developed. The principal state for this phase will be called *acquisition*, and is entered at the top of figure 5.4. An actions decision is then made on the basis of x_1, as shown. If $x_1 = 1$, the appropriate action will be to increment a counter, whose running count is N_1. Notes of the meanings of the various flow-chart symbols are made at the bottom of the chart. If $x_1 = 0$, an actions variable w_1 is inspected. This variable is the output of the magnitude comparator, which compares the counter (N_1) to the data register (N_2). If $w_1 = 1$, $N_1 < N_2$ and the data register should be updated. If $w_1 = 0$, the data register should not be disturbed. This assignment for w_1 has been arbitrarily chosen. In both cases,

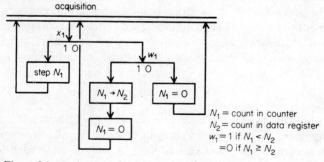

Figure 5.4   *Preliminary flow chart for the acquisition phase of example 5.1.*

the counter should be reset ($N_1 = 0$). This information is added to figure 5.4.

On close examination of the flow chart, however, several snags are discovered.

(1) For $x_1 = 0$, $w_1 = 1$, the N_1 count cannot be both transferred correctly and reset at the same time.

(2) On the second pass through the $x_1 = 0$ condition, w_1 will certainly be at 1, since $N_1 = 0 < N_2$. Hence N_2 will be set to zero during this pass.

(3) There is no provision for flowing from this phase to the transmission phase when it is called for ($x_2 = 1$).

The first and second of these can be overcome by introducing another actions-selector variable, w_2. This variable will be used to guide the flow properly when $x_1 = 0$. On the first pass, let $w_2 = 1$, and compare N_1 to N_2. If $N_1 < N_2$, N_1 will be transferred to the data register. Also, set $w_2 = 0$. On subsequent passes (when $x_1 = w_2 = 0$), reset the counter to zero. A convenient place to set $w_2 = 1$ is in the $x_1 = 1$ flow path. The resulting flow graph is shown in figure 5.5.

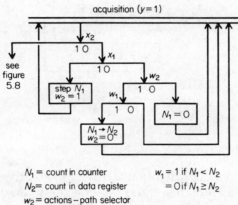

N_1 = count in counter $w_1 = 1$ if $N_1 < N_2$
N_2 = count in data register $= 0$ if $N_1 \geq N_2$
w_2 = actions–path selector

Figure 5.5   *Final flow chart for the acquisition phase of example* 5.1.

The above solution, which overcomes snags 1 and 2, is relatively straightforward. No claims to uniqueness are made, however. There are many other methods that could also be used to overcome these snags. For example, more states could be created. In fact, the w_2 variable may be considered as an additional state variable. The point is that, from the many possibilities that exist, a solution is desired that is (a) simple to implement and (b) simple to describe on a flow chart. The optimal solution in many cases is a matter of personal taste.

The final snag to be overcome is the provision of a means for getting to the transmission phase. Remember that, on every clock pulse, flow takes place from the acquisition state to an action path and then back to the acquisition state. It is only necessary to query the x_2 variable before proceeding to the x_1 decision. The result is shown in figure 5.5. The result of what happens when

$x_2 = 1$ is not known at this stage, and so its output is directed to some subsequent figure. The design of the transmission-phase flow chart is considered next. It would be a good plan to re-use the counter, data register, and magnitude comparator, if possible, to simplify the design. A toggle flip–flop will also be needed. This flip–flop changes state upon the application of a clock pulse. It can be formed by setting $J=1$, $K=1$ in a JK flip–flop, for example. The central idea will be to increment the counter until it reaches N_2. At this point, toggle the flip–flop (whose output will be z) and reset the counter. The design is straightforward, and the resulting flow chart is shown in figure 5.6. Once again the flow from the transmission phase to the acquisition phase has been left for some subsequent figure. The flow graphs for the individual phases are now complete.

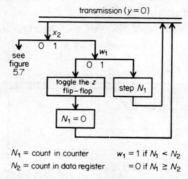

Figure 5.6   *Transmission-phase flow chart for example 5.1.*

5.3.3   *Inter-phase linkages*

The next portion of the design to consider is the linking together of the above two phases. The link from the transmission phase to the acquisition phase will be considered first. The best time to leave the transmission phase is when $x_1 = 0$. If transfer to the acquisition phase was made when $x_1 = 1$, a spurious count would then be recorded. In passing to the acquisition phase it is also necessary to reset the counter and the z and w_2 flip–flops to *zero*, and to set the data register to its maximum value. This latter step will ensure that the data register is started properly (if the data register were set to zero, this value would be recorded as the minimum half-period). The resulting linkage is shown in figure 5.7. Note that two of the paths have been annotated. Labelling such as this can be of real value in understanding the operation of a flow chart at a glance.

Only one possibility for the inter-phase linking has been shown. In general there will be a number of possibilities. New states can be created, as well as forming new w_i actions variables. However, simplicity of design and ease of understanding should again form useful criteria in finding the optimum linkage. The link from the acquisition phase to the transmission phase is

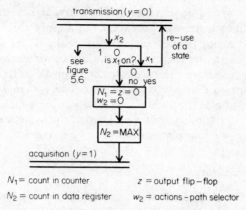

Figure 5.7   *Linkage from the transmission phase to the acquisition phase.*

considered next. The main point to check is whether N_1 should be trans-ferred to N_2 during this transition. A realisation of this linkage is given in figure 5.8. Note that, if N_1 should be transferred to the data register, use is made of the acquisition state again before completing the transition to the transmission state. Figure 5.8 is in many ways similar to figure 5.5.

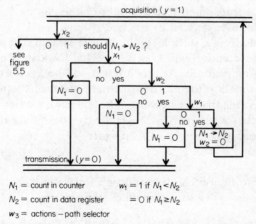

Figure 5.8   *Linkage from the acquisition phase to the transmission phase.*

5.3.4   Initialisation

Consideration is now given to the initialisation or starting-up of the system. It is logical to assume that when the power is switched on the system should proceed from some initial state to the acquisition phase. In general, initialisa-tion is important as it is impossible to predict the values that the flip–flops (and counters) will go to when power is first applied. One must take definite action to ensure that all flip–flops are at their correct starting values. In practice simple starting-up circuits are employed. For example, a series resistance–capacitance network between the supply rails will provide an exponentially increasing (or decreasing) voltage that is sufficient to preset or reset flip–flops at first, and then to have no further influence. For the present example, the starting state can be made the same as the transmission state, under the assumption that $x_2 = 0$ at the start. The system will then make an inter-phase transition to the acquisition state, during which N_1, z and w_2 are reset. If there is a possibility that $x_2 = 1$ when the system is started, then a more complicated starting plan will have to be devised.

5.3.5   *Circuit realisation*

In this example two different states are used in the flow chart. As the states are defined by flip–flops, then, as $2^1 = 2$, one flip–flop will be necessary. If five to eight states had been used, then three flip–flops would be necessary, and so forth. The assignment of binary numbers to the states is called the state-assignment problem. This problem in general is difficult to solve. Some useful guide-lines will be given in section 5.5. For example 5.1, however, only one state flip–flop y is involved, and so there are only two possible state assignments. Recall that at switch-on it is desirable to be in the transmission state. Nearly all flip–flops have a reset facility; some also have a preset facility. Thus the transmission state will be defined by $y = 0$, which leaves the acquisition state to be defined by $y = 1$.

Having decided on a state assignment, the system can now be put into hardware form. Attention will first be directed towards the state-transition combinational-logic circuit. At this point it will be assumed that y is to be stored in a JK flip–flop (details of which are given in chapter 3). Referring to figure 5.7, it is seen that the flow from the transmission ($y = 0$) to acquisition ($y = 1$) state occurs when $x_2 x_1 = 00$. Hence by combining the condition for transfer and the initial state in an AND gate

$$J_y = \bar{x}_1 \bar{x}_2 \bar{y}$$

Next, referring to figure 5.8, the condition for transfer from the acquisition ($y = 1$) to transmission ($y = 0$) state takes place when

$$x_2 x_1 + x_2 \bar{x}_1 \bar{w}_2 + x_2 \bar{x}_1 w_2 \bar{w}_1 = 1$$

This result, when combined with the acquisition state, gives

$$K_y = (x_2 x_1 + x_2 \bar{x}_1 \bar{w}_2 + x_2 \bar{x}_1 w_2 \bar{w}_1) y$$

The circuitry resulting from these equations, together with a starting-up circuit, are shown in figure 5.9. It turns out that due to the nature of the J–K flip–flop the y and $\bar{y}$ variables in the above two equations are not needed in this example. They have been included, however, to illustrate the general technique of determining the state-transitional logic. Map factoring and other combinational-design techniques can be used to modify this logic,

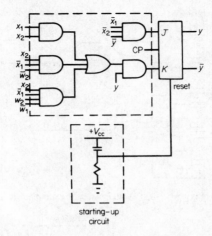

Figure 5.9 *State-transition circuit for example 5.1.*

if desired. As races and hazards are not important, many levels of gates can be employed in this logic block. However, it should be kept in mind that, if there is a direct relationship between the logic and the flow graphs, the increased understanding that results could well outweigh the advantage of a complicated gate arrangement that saves a gate or two.

The actions portion of the system is considered next; figures 5.5, 5.6, 5.7 and 5.8 are the relevant ones. At the start of this design the states should be labelled for clarity. For example, $y = 0$ should be put beside the transmission state. Each actions function is considered in turn. The counter is a useful device to begin with. Referring to figures 5.5 and 5.6, the counter is incremented when

$$y\bar{x}_2 x_1 + \bar{y} x_2 w_1 = 1$$

The counter is reset when

$$y\bar{x}_2 \bar{x}_1 \bar{w}_2 + \bar{y} x_2 \bar{w}_1 + \bar{y} \bar{x}_2 \bar{x}_1 + K_y = 1$$

(the left-hand terms come from figures 5.5, 5.6, 5.7 and 5.8, respectively), where K_y is the flip–flop input that was derived earlier. The resulting circuitry for the counter is shown in figure 5.10. The w_2 variable is considered next. This variable is a flip–flop (a set–reset asynchronous flip–flop, say) that, from figures 5.5 and 5.7 has the excitation equations

$$\text{SET} = y\bar{x}_2 x_1$$
$$\text{RESET} = y\bar{x}_2 \bar{x}_1 w_2 w_1 + \bar{y} \bar{x}_2 \bar{x}_1$$

The logic for this flip–flop is depicted in figure 5.11.

The w_1 variable is the output of the magnitude comparator, which compares the outputs of the counter and the data register. The data register

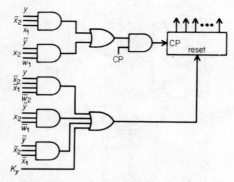

Figure 5.10   *Counter logic for example* 5.1.

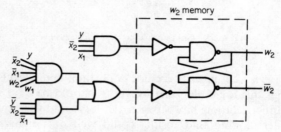

Figure 5.11   *Logic for w_2 of example* 5.1.

accepts fresh information from the counter when

$$y\bar{x}_2\bar{x}_1 w_2 w_1 + yx_2\bar{x}_1 w_2 w_1 \ (= y\bar{x}_1 w_2 w_1) = 1$$

It is set to its maximum value when

$$\bar{y}\bar{x}_2\bar{x}_1 = 1$$

The above components are shown in figure 5.12.

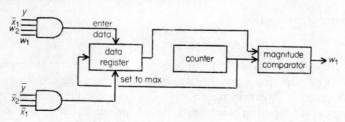

Figure 5.12   *Additional logic for example* 5.1.

Finally, the output flip–flop z is considered. It is reset when

$$\bar{y}\bar{x}_2\bar{x}_1 = 1$$

and is toggled when

$$\bar{y}x_2\bar{w}_1 = 1$$

Circuit details are shown in figure 5.13. The reset facility (which is normally asynchronous) on the z flip–flop has been used to simplify the design. This means that the z reset will start as soon as $\bar{y}\bar{x}_2\bar{x}_1 = 1$. If synchronous reset is desired, additional clock gating will be necessary. In general, a feature of this design method is that a mixture of synchronous and asynchronous control can be used.

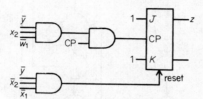

Figure 5.13 *Output logic for example* 5.1.

It will be noticed in this example that circuit duplication exists. For example, the expression $y\bar{x}_2x_1$ appears in figures 5.10 and 5.11. It should be mentioned that if the system is to be built on a modular basis the duplication of gates on separate boards can be desirable. Fewer inter-board connections may be required. In other cases, one of the gates may be eliminated.

Although the circuit realisation appears to be complicated, it bears a direct relation to the flow charts and, in fact, is derived by inspection of the charts. In practice the realisation can be quickly obtained, which is another advantage of the flow-chart technique. If the system should develop a malfunction, the fault can be quickly located with the aid of the related flow charts. This fault-finding aspect again emphasises the need of producing a well-labelled flow chart, the operation of which is simple to follow.

5.4 Pictorial aids

5.4.1 State-transition diagram

In some cases the specification will not require actions to be performed in moving between the different states. In this situation the flow chart reduces to a state-transition diagram of the type described by Mealy (1955). This diagram (which is similar to the one developed in chapter 4) is introduced by way of the following example.

Example 5.2

Design a synchronous system that will recognise the input sequence $x = 0$, 0, 1, 0. The output z is to become *one* when the final *zero* of the above pattern appears.

Solution The procedure starts with identifying the flow for a correct input pattern, as shown in figure 5.14a. The next step in the solution is to fill in the

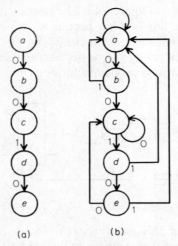

(a)   (b)

Figure 5.14   *State-transition diagram for example* 5.2.

state-transition diagram for the other possibilities, with the idea of re-using states whenever possible. If the input should be at *one* when moving from state *a*, for example, the flow can be conveniently returned to state *a*. Note that the solution in figure 5.14b allows for pattern overlap. That is, if the input pattern should be $x = 0, 0, 1, \underline{0}, 0, 1, \underline{0}$, an output of *one* will occur at each of the underlined input bits. In this example re-use is made of existing states in considering all possible transitions, so that only five states are necessary. Three flip–flops will be required.

This example can also be tackled with the flow-chart method. Various shift registers are available, together with suitable binary comparators. A suitable flow chart is shown in figure 5.15. The shift register, which is in the actions block, provides all the necessary memory requirements. A system-state memory is not needed as there is only one system state. In the above example, the shift-register approach will require two more flip–flops than a formal-synthesis approach (which uses four states). In fact, if one takes into account

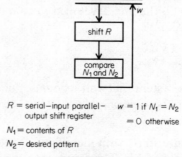

R = serial–input parallel– w = 1 if $N_1 = N_2$
     output shift register
                                   = 0 otherwise
N_1 = contents of R
N_2 = desired pattern

Figure 5.15   *Flow chart for example* 5.2.

the ease of design and the ease of subsequent understanding of system operation, the shift-register solution is more satisfactory. This solution is also very flexible; the detected code can be easily modified.

5.4.2 State table

An alternative to the state-transition diagram is the state table as described in section 4.4.2. Either representation can be used for designing synchronous systems having no actions. The state table can be viewed as a tabular list of transitions between states. An example of such a table for a synchronous system is given in figure 5.16.

$x_1x_2 =$	00	01	11	10	00	01	11	10
s^v		s^{v+1}				z^v		
s_1	s_1	s_2	s_2	s_3	1	0	0	0
s_2	s_3	s_2	s_3	s_3	0	0	0	0
s_3	s_1	s_2	—	s_4	0	0	—	0
s_4	s_3	s_1	s_2	s_4	0	0	0	0

Figure 5.16 *Synchronous state table.*

The small letter s stands for state. The superscript v indicates the clock-pulse number. Here the $v+1$ clock pulse directly follows the v clock pulse, where v is an arbitrary integer. The states are assumed to exist during the time intervals D_1 and A_2 in figure 5.2. The subscripts are used to identify each state that exists in the system. In this example there are four distinct states: s_1, s_2, s_3 and s_4. The input variables x_1 and x_2 are given a Gray-code ordering, in anticipation of future table factoring. By the use of figure 5.16 one can identify the state transitions that will occur from one clock pulse to the next. For example, if the system is in state s_1 at the arrival of the tenth clock pulse, and during this time $x_1x_2 = 01$, then at the arrival of the eleventh clock pulse the system will be in state s_2. Note that the system does not necessarily change state from clock pulse to clock pulse. For example, if state s_1 should exist and x_1x_2 should stay at 00, the system will remain in state s_1 from clock pulse to clock pulse. In figure 5.16 every next state is a stable state; there are no unstable states in this figure. Also multiple-input transitions are allowed from one clock pulse to the next, so that the fundamental-mode assumption of asynchronous circuits is not needed. In figure 5.16 the blank that occurs in the third row indicates a don't-care condition. If one or more don't cares occur in the table it is referred to as an incompletely specified state table. A don't care will arise when the designer knows that, for some reason, a certain input is impossible when the system is in a certain state. Because of device failures, noise, etc., there is always a possibility that the forbidden input combination will arise sooner or later. When it does, the designer might well want to know about it. To do so, the system could be specified to transfer to an emergency state, given the present state and the forbidden input. This emergency state would then provide some indication (flashing a light, or stepping an error counter, for example) that all was not well.

An advantage of the state table is that it forces the designer to consider all input and present-state combinations, in that every entry in the table should be filled in. A disadvantage of this table is that the designer loses a pictorial view of the flow that exists between the states in the system. For example, consider the state-transition diagram in figure 5.17. The input and output

Figure 5.17   *State-transition diagram example.*

variables have been omitted from this figure for the sake of clarity. Suppose, however, that the input patterns occur at random. Then, regardless of the starting state, it is clear that sooner or later the system will flow to the three states d, e and f as a consequence of the transition from c to e, which has no corresponding return path. The group of states a, b and c can only function as transient states. This knowledge is gained quite easily from the state-transition diagram, but it would be difficult to extract from a state table. Of course the designer can prepare both a state-transition diagram and a state table, if desired. There is obviously a one-to-one correspondence between the two representations. Given a state-transition diagram, its state table can be easily derived, and vice versa.

On the state table in figure 5.16 the system output z is also displayed. The don't-care next-state blank gives rise to a corresponding blank in the output section of the table. There can of course be more than one output signal. Only one output variable is shown for the sake of clarity.

5.4.3   *Equivalent-state detection*

It sometimes (but not very often) occurs that more states are used in the synchronous state table than are necessary. Recall that for this case all states are stable. For example, if two rows in the state table of figure 5.16 are as shown in figure 5.18, then it is obvious that s_1 is equivalent to s_2 and so one of these rows can be removed. If the second row is removed, then s_2 can be replaced by s_1 in the remaining table. It is beneficial to remove these redundant rows since the number of states is thereby reduced, which can sometimes lead to a reduction in state variables. Recall that n state variables can accommodate 2^n rows or states in the state table.

Matters are usually not as straightforward as in the above example. Consider the two rows of figure 5.19, whose outputs are the same. Is s_1

s^v		s^{v+1}			z^v			
s_1	s_4	s_5	s_4	s_5	1	0	1	1
s_2	s_4	s_5	s_4	s_5	1	0	1	1

Figure 5.18   *State table. Example of equivalence.*

s^v		s^{v+1}		
s_1	s_4	s_5	s_4	s_5
s_2	s_4	s_5	s_6	s_5

Figure 5.19   *State table. Example of possible equivalence*

equivalent to s_2 $(s_1 = s_2)$? It is seen that they are only equivalent if s_4 is equivalent to s_6. Suppose also that the two rows of figure 5.20 appear (having identical outputs) in the same table. Is $s_4 = s_6$? Again, the answer is yes only if $s_1 = s_2$. Here the argument has gone full circle. That is, $s_1 = s_2$ if $s_4 = s_6$ and $s_4 = s_6$ if $s_1 = s_2$. At this point, two assumptions can be made. First, $s_1 = s_2$.

s^v		s^{v+1}		
s_4	s_1	s_3	s_3	s_3
s_6	s_2	s_3	s_3	s_3

Figure 5.20   *State table. Example of possible equivalence.*

This assumption would result in $s_4 = s_6$, and so states s_2 and s_6 could then be removed from the table. Second, $s_1 \neq s_2$. This assumption would then imply that $s_4 \neq s_6$, and so all four states would remain in the table. Since the object here is to use as few states as possible, the first assumption $(s_1 = s_2)$ is preferable. Two states then become redundant, and can be removed.

Checking for equivalent states can be facilitated through the use of the implication chart method of section 4.4.3.

Recall that two states are not equivalent if their outputs differ. Outputs should be checked first to establish possible equivalences. A different implication chart is then formed for each output requirement, and the procedure in section 4.4.3 is followed. Finally, it is generally worth while to relabel all non-redundant states at the end of the procedure in ascending order (for example, A, B, C, D). A non-redundant state table is them formed using these relabelled states.

A technique for eliminating redundant states has been given. In theory there is no reason why a designer could not use ten or a hundred redundant states in the initial specification, and then remove these states with the above technique. There are two reasons why this approach is not good. Firstly, the use of many redundant states is time-consuming, in both the initial specification and the removal of equivalent states. Secondly, in dealing with a large number of states there is a greater chance for human error to occur. If a non-redundant state were accidently eliminated, for example, it could prove very time-consuming to discover the error. For these reasons it is a good plan to re-use states when forming the initial specification. If the states are re-used judiciously, then, when checking for redundant states, only a few tests need be made. Recalling that states with different outputs cannot be equivalent, the testing can usually be done by inspection. An example of state re-use was given in example 5.2.

5.5   State assignment

In the preceeding sections of this chapter the concept of a state has been discussed, as well as a technique for removing redundant states. The next matter to consider is how to assign or label the states with the system state variables y_i. These variables are normally flip–flop outputs and so, to change the system from one state to another, the states of the necessary flip–flops are changed.

With n internal variables there are 2^n different bit patterns (or words). There are 2^n different possibilities of assigning a word to the first state, $2^n - 1$ ways of assigning one of the remaining words to the second state, and so on. There are certain redundancies, however, which lower the possibilities. Details are given in section 4.4.6 and, in particular, figure 4.28. The point to observe is that the number of possible assignments grows rapidly with n. From this wide variety of permutations, the state-assignment problem is to establish some suitable criterion that will reject all but a few (or possibly one) of these assignments, and then to determine the assignment that 'best' meets this criterion.

In the past the criterion has usually been firstly to minimise the number of flip–flops or memory required, and secondly to minimise the combinational logic (or number of gates). In some cases this criterion will also minimise cost. For example, if the gates and flip–flops have to be built using discrete components then the cost will be relatively small with this criterion.

With the integrated circuits of today, however, a different criterion should be established. For example, combinational-logic gates of identical type are usually packaged together in a chip. If one NOR gate and one NAND gate are required, two chips might be necessary, whereas, by judiciously redesigning the logic, it might be possible to use three NAND gates in one chip. With regard to memory storage it might occur that either three discrete flip–flops (8 states) or a 5-bit shift register are required. The former would require two chips (assuming dual-flip–flop chips), whereas the latter can be obtained in one chip. As the reader can see, emphasis is swinging towards the use of a minimal number of chips in the design. This criterion is important in both minimising cost and improving reliability. The cost is also minimised because of the associated printed-circuit-board cost. Surface area on the board can cost several times more than the chip itself, if one takes into account the design time, manufacturing costs, etc. Hence the use of as few chips as possible will minimise this cost. In many cases the drain current from the power supply is also minimal when the fewest number of chips is used. From a reliability point of view, the fewer the number of soldered joints the better. When one takes into account that with a 14-pin chip the power-supply connections represent over 14 per cent of the total connections it is seen that the smaller the number of chips the better.

From the above discussion it would appear that the optimal design strategy is to use as few chips as possible, and this is indeed the case in many circumstances. For example, in designing a system that will be produced by the thousands or millions, elimination of even one chip can result in substantial cost reduction. The operation of the system does not have to be well understood or easy to follow. If a fault occurs it is simply necessary to remove (or unplug) the system and to put a good one in its place. Usually systems of this nature will be built on a modular or plug-in basis, and so are relatively easy to 'fix'. The faulty system is then sent back to the factory for repair or, at times, simply discarded. The repair department will have the skill and experience necessary to effect the repair.

The state-assignment problem for this task is extremely important, since it affects not only the amount of logic required for the internal memory but also the amount required for the output circuitry. Up to this time there is no general method that, when used, will guarantee an overall minimum-gate design. In practice a variety of techniques are used. The reader is referred to Hill and Peterson (1968) for a discussion of these techniques.

The other case of interest is the situation where one or possibly a few of the units are required. This case can arise in designing in-house special equipment. For example, an automated test system may be required, which is not complicated enough to warrant on-line computer control, but nevertheless employs sequential logic. In the state assignments for such systems the human factor becomes very important. If one can determine a state assignment that is simple to understand and follow, the following items will be optimised with respect to human factors

(1) design time
(2) design modification
(3) system description in reports
(4) repair time.

The system will *not* usually be optimised for minimal chip count and hence for chip cost under the above assignment. But as far as the *total* design cost is concerned, this assignment is usually optimal. With respect to item 1, for example, the cost of one man-hour exceeds the cost of most chips. With regard to item 2, it is often required to modify a system to provide new capabilities. If the state assignment is simple, it should be simple to modify. On the other hand, if the assignment has been fairly complicated, a completely new assignment might be necessary for the modification. There is also the problem of communicating the system mode of operation to other people. A simple state assignment will facilitate this task, as item 3 suggests. Finally, there is the debugging and repair aspect of item 4. The need for a simple state assignment here should be self-evident. The means for meeting these criteria is implicit in the flow-chart method.

To determine a suitable state assignment for human use, consideration should be given first to the various phases of operation. These phases should be allocated inter-phase flip–flops that are used only for phase control. The operation of the system within a phase should be determined by a different set of flip–flops, which are re-used for every phase. By this means it is easy to determine the system state at each step. The phase flip–flops give only phase information, and the other flip–flops give only intraphase information. This two-level approach is especially useful in describing the system, because a different flow graph can be prepared for each phase. The phase flow graph can be very detailed, and will enable the reader to concentrate on just one phase of operation rather than on the complete system. A phase-linkage flow graph can also be prepared, in which each phase is treated like a state on the phase flow graph. For instance, in example 5.1 there are two phases, and so one inter-phase flip–flop will be required. No intraphase flip–flops are required. In general, if there are p phases and q (maximum) intraphase states, then $M + N$ flip–flops will be required, where

$$2^{M-1} < p \leqslant 2^M$$
$$2^{N-1} < q \leqslant 2^N$$

This state-assignment technique meets all of the criteria listed earlier. It is simple to develop—in many cases the design can be completed (the states labelled) on the system flow chart itself.

5.6   Examples

This section contains both worked and unworked examples that will illustrate many of the salient features and some extensions of this chapter.

Example 5.3

The logic system of figure 5.21 is to be constructed only once. It might be necessary to add more phases at a future date. Find a state assignment that is suitable for this system.

Solution   In figure 5.21 it is seen that the flow chart has already been partitioned into phases, as indicated by the letters. Note that phases to start up the system properly and to halt the system are included. For this problem two classes of flip–flops will be used. The inter-phase flip–flops M_i will direct the flow from phase to phase, whilst the intraphase flip–flops N_i will be used to control the flow in a particular phase. It is seen that phases A and B have the most states (nine), which indicates that four intraphase flip–flops $N_1 N_2 N_3 N_4$

will be required. These flip–flops are re-used in each phase. Therefore, provision must be made on the flow chart to reset the N_i when moving from one phase to another. This in turn implies that the states A_1, B_1 and D_1 will be partially defined by $N_1N_2N_3N_4 = 0000$. The assignment of the N_i in a phase is fairly arbitrary. An often-used rule of thumb is to arrange adjacent states within a phase such that only one of the N_i changes, wherever possible. This strategy will tend to minimise the amount of logic (and wires) needed for proper control of the N_i flip–flops. Thus, B_2 and B_3 might be (partially) defined by $N_1N_2N_3N_4 = 0001$ and 0010.

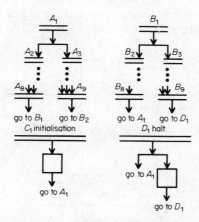

Figure 5.21   *Flow chart for example 5.3.*

The inter-phase flip–flops M_i are considered next. To allow for future expansion, three M_i will be used. Then four surplus phases will exist. The obvious M_i to use for C_1 is $M_1M_2M_3 = 000$, which can be accomplished quite simply at switch-on by resetting all M_i. The assignments for the other phases are arbitrary, except that the logic is reduced if only one M_i is changed in passing from one phase to another.

In this example every state is identified by the pattern of *ones* and *zeros* in the expression $M_1M_2M_3N_1N_2N_3N_4$. In figure 5.21 20 states are included. With seven variables it is possible to have 128 states, and so a usage efficiency of 16 per cent exists. This is fairly low, but future expansion is possible, and the action of the circuit is quite simple to understand.

In general the numbering of the states can be likened to the numbering of sections and chapters of a book. In fact, an instruction manual could be written such that each chapter is named after a phase, and the sections of each chapter give a state-by-state description of operation within the phase. For a very large system, a third level of memory variables can be employed. These variables would correspond to the names of books, with each book having the correspondences given above.

Example 5.4

Determine the flow map for a synchronous system whose output will go to *one* when any of the sequential input patterns 0101, 0011, 1001 and 1011 occur. The system is to go to a rest state for all other input patterns and for all subsequent inputs.

Solution The solution is started by laying out a state-transition diagram for the input pattern 0101, as shown in figure 5.22a. Here the successful transition flows from left to right. The circles represent states, and the boxes represent final rest states. Three boxes can immediately be entered, as shown. The next pattern 0011 is added to this diagram, by making use of one of the states for the previous pattern. The result is shown in figure 5.22b. In a similar manner

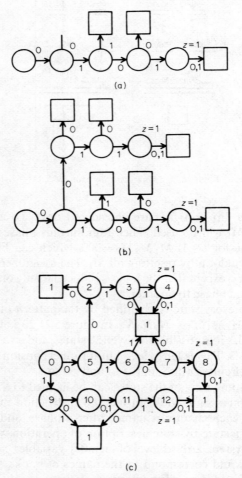

Figure 5.22   *Transition diagrams for example 5.4.*

the other two patterns are added, the result being figure 5.22c. The states have also been labelled in this figure.

The next step is to form a flow map to check for redundancy. The flow map corresponding to figure 5.22c is shown in figure 5.23a. The map immediately shows that s_4, s_8 and s_{12} are equivalent. The flow map then indicates that s_3, s_7 and s_{11} are equivalent. Replacing s_8 and s_{12} by s_4 and s_7 and s_{11} by s_3,

(a)

s^v	s^{v+1}		z^v
	$x=0$	1	
s_0	s_5	s_9	
s_1	s_1	s_1	
s_2	s_1	s_3	
s_3	s_1	s_4	
s_4	s_1	s_1	1
s_5	s_2	s_6	
s_6	s_7	s_1	
s_7	s_1	s_8	
s_8	s_1	s_1	1
s_9	s_{10}	s_1	
s_{10}	s_{11}	s_{11}	
s_{11}	s_1	s_{12}	
s_{12}	s_1	s_1	1

(b)

s^v	s^{v+1}		z^v
	$x=0$	1	
s_0	s_5	s_9	
s_1	s_1	s_1	
s_2	s_1	s_3	
s_3	s_1	s_4	
s_4	s_1	s_1	1
s_5	s_2	s_6	
s_6	s_3	s_1	
s_9	s_{10}	s_1	
s_{10}	s_3	s_3	

Figure 5.23 *Flow maps for example 5.4.*

the reduced map of figure 5.23b is obtained. This map can be checked for redundancy by first comparing s_0 and s_1, s_2, ..., s_9, then comparing s_1 and s_2, s_3, ..., s_9, and so on. Recalling that s_1 and s_4 are distinct, it is readily established that there are no more redundancies. For example, $s_0 = s_5$ only if $s_2 = s_5$ and $s_9 = s_6$. But $s_2 \neq s_5$ since $s_1 \neq s_2$. Thus, $s_0 \neq s_5$. Finally, a reduced transition diagram can be drawn from this map, as indicated in figure 5.24. The states can be renumbered with consecutive integers, if desired.

This example illustrates the interplay between the state-transition diagram and the flow map. In checking the flow map for redundancies it is easily established that there are $(n-1)(n-2)\ldots(2)(1)=(n-1)!$ checks to make. Therefore it is desirable to remove most of the redundancy on the initial

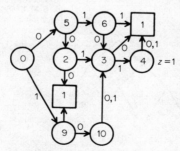

Figure 5.24 *Final state-transition diagram for example 5.4.*

state-transition diagram. The remaining redundancy can then easily be found from the corresponding flow map.

Example 5.5

Find a state assignment for the previous example such that transitions between adjacent states (as shown in figure 5.24) involve the change of only one of the four variables, except for transitions to the rest state s_1. (Hint: Place the states on a four-variable Karnaugh map.)

Example 5.6

In a certain flow chart, states are used for two purposes. The first set of states A_i is used for making binary decisions. A second set of states $B_1, B_2, \ldots, B_6$ is used for various actions. The outcome of several A_i leads to one of the B_i. The probability of occurrence p_i of each B_i is also known. In this example, $p_1 = 0.08$, $p_2 = 0.14$, $p_3 = 0.16$, $p_4 = 0.18$, $p_5 = 0.20$ and $p_6 = 0.24$. Find an arrangement of the A_i such that the operation time of the system is minimised.

Solution   The probability numbers indicate how often a certain path will be followed. For example, B_1 will occur eight times for every hundred starts. Let N_i equal the number of states visited before reaching a B_i. Then the problem is to arrange the A_i such that the sum $\Sigma N_i p_i$ is minimised. This problem has been dealt with by Knuth (1972). One interpretation of his solution is as follows. The probability numbers (scaled by 100) are put in a row, as shown in figure 5.25a, in ascending order. The two lowest numbers (8 and 14) are merged together, as shown by the slanted lines below the numbers. The sum

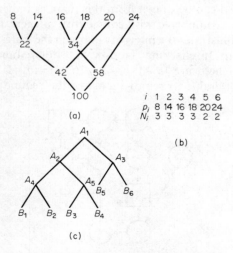

Figure 5.25   *Trees used in example 5.6.*

of these numbers (22) is placed at the focal point of the slanted lines. This sum is now treated as a single row number. The result of this operation is to reduce the number of row numbers by one. The procedure is repeated over and over, until only a single number remains. If at a particular stage the lowest number occurs three or more times, arbitrarily pick two of the entries and proceed as above.

The resulting tree can be put in a more presentable form as follows. A list is made of the probability numbers and their path lengths N_i, as displayed in figure 5.25b. In general N_i should always decrease in moving from left to right. The final tree is constructed by moving through this list from right to left. B_6 for example has a path length of two, and so can immediately be drawn as shown in figure 5.25c. The other B_i are entered in descending order, as indicated. At each node in the tree a different A_i is entered to complete the solution.

Example 5.7

Using the technique of example 5.6, find a minimal tree for $p_1 = 0.02$, $p_2 = 0.04$, $p_3 = 0.06$, $p_4 = 0.08$, $p_5 = 0.12$, $p_6 = 0.14$, $p_7 = 0.16$, $p_8 = 0.18$ and $p_9 = 0.20$.

Example 5.8

A serial data message is sent over a data channel for control purposes. The message consists of the pattern 01 repeated eight times (for synchronisation purposes), the identification word 0011, and finally a control word $00 \times \times$ (each $\times$ is either a *zero* or a *one*), which is used to initiate one of four actions. A clock pulse synchronised to the input data is also assumed to exist. An alarm is to be turned on if any deviation from the above sequence occurs. Design a system that will meet this specification.

References

Hill, F. J., and Peterson, G. R., *Introduction to Switching Theory and Logical Design* (Wiley, New York, 1968).

Knuth, D. E., *The Art of Computer Programming, Volume 1 Fundamental Algorithms* (Addison-Wesley, Reading, Mass., 1972).

Mealy, G. H., 'A Method for Synthesising Sequential Circuits', *Bell Syst. tech. J.*, 34 (1955) pp. 1045–79.

6

Practical Design Considerations

6.1   Initial specification

In system design a good starting point is to regard the system as a black box. The purpose of the black box is to accept certain inputs, process them, and then to produce certain outputs. In this sense the system performs a *function*, similar to the case where $g(x)$ is a function of x in mathematical analysis. Some initial specifications are literally a listing of inputs on the left side of a sheet of paper, and the corresponding outputs on the right side. Or perhaps a collection of waveform-timing diagrams are to be realised. At this stage one should not be concerned with the mass of detail that will follow later. The important thing is to get the *functional* description right.

Consideration should then be given to the front-panel design and operator controls. One should decide what to display, the type of display (for example, analogue or digital) and the accuracy required. The trend in operator controls is to reduce their number to a minimum. For example, a fine frequency-tuning control could be eliminated by a phase-locked loop in the system. The subject of human factors should also enter at this point. Items such as the size, colour and brightness of the displays, the types and grouping of the controls, and how often displays should be updated can be decided. If possible, a drawing of the front panel should be produced. Too often this part of the design is left until *after* the detailed circuit design. Then front-panel modifications add to the design cost.

6.2   Detailed design

Having achieved a functional specification, the next step is to narrow down the enormous number of realisation possibilities. The first objective is to decide on what technologies will be used. From the initial specifications one can say roughly what speeds, power consumption, volume and working environments will be necessary. These considerations will in turn suggest certain technologies. For example, if high speed is a prime requirement then emitter-coupled logic rather than fluidic logic should be used. Usually the appropriate technologies can be determined quite rapidly.

The next step is to produce a block diagram for the system. Enough blocks should be included so that each block will require only a moderate design effort. Each block should be identified with a word description of its function and the technology that will be used. Typical examples are 'voltage-controlled oscillator, discrete components' and 'microprocessor, integrated circuits'.

The design of each block is then considered in turn. Manufacturers' catalogues should be used here to locate any special-purpose devices that would prove useful. For example, a medium-scale-integration integrated circuit that performs an up–down counter function is readily available and can be used for a variety of applications. When looking at device specifications, it is important to define the use to which the device is put. For example, if the system is to be mass-produced and speed is important, then the worst-case propagation delay time should be used, rather than the 'typical' delay time. All data sheets should be critically examined, including footnotes! Certain types of 'averaging' can take place on data sheets. For example, a one-to-zero transition may be significantly slower than a zero-to-one transition. On the data sheet, however, one may find only an 'average' transition time. Again it is the worst case that is usually important, not the average case. If the system is to be mass-produced, the question of second sourcing of the special-purpose devices ought to be investigated. Some of these devices, and indeed whole logic families, have been known to suddenly disappear.

Having identified the special-purpose devices, the next step is to produce a flow-chart for each block (if applicable), as described in chapter 5. In establishing a block flow chart, the designer usually has the relevant catalogues at hand, and might require two or three design iterations for the block to make it as efficient as possible. The importance of labelling states with meaningful phrases cannot be over-emphasised at this stage. After the individual blocks have been designed, a block-linkage diagram should be formulated, taking care to include any buffering that might be needed between the blocks.

The next objective is a detailed schematic diagram of the system. Before this diagram can be produced, the states on the flow charts must be assigned binary numbers. Chapter 5 gives some guidelines for this state-assignment problem. After the states have been assigned, a logic diagram for the system should be produced. Using the techniques outlined in chapter 2, one can minimise the combinational logic required, including the number of inverters. Finally, a detailed schematic (or wiring) diagram should be made.

6.3   Prototype development

The first objective is to make a breadboard unit that works. It is usually possible to do this by developing the breadboard block by block. Each block is constructed and tested. If the block does not perform satisfactorily, some design modification will be necessary. When the blocks are working as intended, they should be connected together to check that the overall system performance meets the initial specification. Again, some design modification might be required here.

The next objective is to add any needed input or output buffering to the system. For example, if the system output is to be transmitted through a noisy environment, a line driver could be added to the output. Or if a digital input is known to have noise errors, an error-detection (or error-correction) code might be used for the input word. Suitable decoding will then be necessary at the system input. These additional circuits should be added to the bread-board. Then the system as a whole is tested to ensure proper working.

At this point it is convenient to make a drawing of the final circuit. Elements should be labelled with their numerical values (for example, 4.7 kΩ) but *not* with symbolic labels (for example, $R17$). These labels will be assigned after the printed-circuit boards have been produced.

6.4   Printed-circuit boards

The transfer of the breadboard design into printed-circuit form is advantageous when more than two or three units of the system are required. The repeatability of printed circuits is excellent. Therefore, if one unit works, subsequent units should work as well. The other outstanding advantage of printed circuits is that wiring errors are reduced to a minimum. This is not the case with the point-to-point wire-wrap technique, for example, unless the procedure is automated.

In the design of a printed-circuit board, the first step is to arrange the components (including board connectors) on the board so that the wire connections between the components will be reasonably short, subject to the following constraints.

(1)   All integrated circuits should point in the same direction (preferably a direction parallel to an edge).
(2)   All other components should be parallel to or perpendicular to the direction of the integrated circuits.
(3)   The sum of all wire lengths should be as small as possible.

Condition 1 is necessary in order to prevent component-insertion errors and subsequent testing errors. Condition 2 allows a grid design system to be used. The reason for condition 3 is to reduce mutual coupling and crossovers to a minimum. Associated with 3 is the consideration of crossovers and what sort of board to use. In some cases a board with printed wires on only one side is sufficient. If a few crossovers are needed, these can be provided with wire links, which are inserted just like the other components. If the number of links on a single-sided board is large, then a double-sided board (with plated-through holes) should be considered. If the double-sided-board solution still requires many wire links, then a multilayer-board solution ought to be examined. There are many trade-offs in the selection of the best type of board. The reader is referred to Dukes (1961), Coombs (1967) and Ross (1975) for further details. The above design is usually done by trial and error, but can also be solved by a computer-aided design programme.

In board fabrication the essential rule for successful results is cleanliness. Never touch unprotected surfaces with bare fingers, make certain that surfaces have been rinsed well, and so forth. If difficulties arise, the reader is referred to the troubleshooting book on printed-circuit design by Shemilt (1974). If the board is to be used for a one-off application, it is likely that modifications, especially additions, will be desirable later on. Accordingly, it is prudent to leave some spare space on the board for this purpose. The board layout should also take into account connections to and from the board, as previously mentioned. A popular type of connector is the edge connector, which in addition provides mechanical support for the board. Another widespread technique is the integrated-circuit socket plug. The plug is connected to ribbon cable (which consists of a number of coplanar wires bonded together), and is designed to fit directly into an integrated-circuit socket. These input–output sockets can be located at different board locations to ease board wiring requirements. Consideration should also be given to the provision of test points on the board, for either manual testing or ATE (automatic-test-equipment) testing. In the latter case a special multiterminal connector could prove useful.

In the provision of power supplies to the boards, two general points can be made. First, it is essential to have a good earth. Some companies use thick wire braid for the earth lead from the power supplies to the board. If this lead does not have a low enough resistance to both d.c. *and* a.c. signals, all sorts of spurious coupling can occur. Next, it is becoming increasingly popular to use on-board voltage regulators. These regulators are now available in integrated-circuit form at low cost. The idea is then to supply unregulated d.c. voltage to all board regulators. The same unregulated supply can also be used for driving lights, relays, motors and so forth.

After the boards have been designed and constructed, they are tested. Since the wire layout is different from that of the breadboard prototype, there is no guarantee that the system will perform as before. It might be necessary to compensate for speed effects with the addition of some capacitance at certain points in the system. These speed effects (for example, wire propagation delay times) will have more influence on asynchronous than on synchronous systems, as discussed in chapters 4 and 5. If unwanted oscillation occurs, it can often be stopped by the addition of ferrite beads to the board, especially on the power-distribution printed wires. These beads have minimal effect at d.c., but are quite lossy at high a.c. frequencies.

When the board is working satisfactorily, the components can be given symbolic labels as follows (for resistors). The resistor leftmost and upmost (as viewed from the component side) is given the label $R1$. The board is then scanned in the same manner as reading a book, and the scanned resistors given the consecutively increasing labels $R2$, $R3$,... The other types of component are labelled in a similar manner. It is also standard practice to *print* the symbolic label next to the component on the board. The outstanding advantage of this scheme is that components can be quickly located. The symbolic labels are then added to the schematic diagram.

6.5   Testing and documentation

If many units of a system are to be produced, consideration should be given to a schedule of tests at the completion of assembly of each unit. This schedule is usually a compromise, since exhaustive testing of all input possibilities (including front-panel controls) usually takes too long, but each control and input line should be tested at least once to make sure it has been properly connected. A greater range of input possibilities can be tested with automatic test equipment (in a given time). Since this equipment tends to be costly, it is only justified if a reasonably large number of units are produced.

In some cases the testing can be performed quite simply by a comparison technique. The tester (man or machine) has a 'standard' unit that is known to work satisfactorily and a schedule of tests. The output of the unit under test is compared to the standard, the difference being required to lie within stated bounds (as given in the schedule of tests).

In addition to a schedule of tests, an operation manual should also be written. This manual should contain enough detail so that component errors can be detected. The flow charts previously developed for the blocks can be used to advantage, and possibly be included in the manual itself. If this is done, the states ought to be labelled symbolically whenever possible. For example, some typical labels are 'starting-up', 'test for overflow' and 'error has occurred'. A clearly written operation manual is important for both the system user and the repairman. Therefore, sufficient time should be allocated for this task.

6.6   Conclusion

The above material is not intended as an intensive coverage of the design procedure. Rather it is to point out some salient features that have been gathered from practical experience. If these features (and their order of development) are kept in mind, the amount of design time and effort will be kept to a minimum.

References

Coombs, C. F., *Printed Circuits Handbook* (McGraw-Hill, New York, 1967).
Dukes, J. M. C., *Printed Circuits, Their Design and Application* (Macdonald, London, 1961).
Ross, M., *Modern Circuit Technology* (Portcullis, London, 1975).
Shemilt, H. R., *Printed Circuit Troubleshooting* (Electrochemical Publications, Glasgow, 1974).

Further Reading

Friedman, A. D., and Menon, P. R., *Theory and Design of Switching Circuits* (Computer Science Press, 1975).

Appendix Logic Symbols

A wide variety of different symbols has been used to represent logic elements. In this book American MIL standard symbols are used; these are shown in the first column of figure A.1, which also shows the equivalent British Standard symbols. The MIL symbols have the advantage of not requiring any character within the outline to identify the gate function; this space is therefore available to the designer for indication of gate type and/or location.

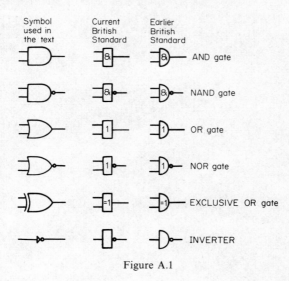

Figure A.1

Index

142